OSTEOPOROSIS
The Clinician's Guide to Diagnosis and Management

First Edition

OSTEOPOROSIS
The Clinician's Guide to Diagnosis and Management

First Edition

Andrew J. Cozadd, PA-C
Orthopedic Surgery
Minneapolis, Minnesota

OSTEOPOROSIS
The Clinician's Guide to Diagnosis and Management
First Edition

© 2016 Norway Street Press

All rights reserved. No part of this publication may be reproduced, stored in a retrieval system, or transmitted, in any form or by any means— electronic, mechanical, photocopying, recording, or otherwise— without written permission from the author.

The contents of this work are intended to further scientific and medical understanding and are not intended and should not be relied upon as recommending or promoting a specific method, diagnosis, or treatment by health science practitioners for any particular patient. Care has been taken to confirm the accuracy of the information presented and to describe generally accepted practices. However, the author is not responsible for errors or omissions or for any consequences from application of the information in this book and makes no warranty, expressed or implied, with respect to the currency, completeness, or accuracy of the contents of this publication. Application of the information in a particular situation remains the professional responsibility of the practitioner.

The author has exerted every effort to ensure that drug selection and dosage set forth in this text are in accordance with current recommendations and practice at the time of publication. However, in view of ongoing research, changes in government regulations, and the constant flow of information relating to drug therapy and drug reactions, the reader is urged to check the package insert for each drug for any change in indications and dosage and for added warnings and precautions. This is particularly important when the recommended agent is a new or infrequently employed drug.

Table of Contents

Preface...1
General Information..2
Screening for Osteoporosis...5
Assessment of Fracture Risk...11
Secondary Osteoporosis: Medical Conditions..........................13
Secondary Osteoporosis: Medications.....................................19
Osteoporosis in Men..22
Osteoporosis in Premenopausal Women.................................26
Osteoporosis in Postmenopausal Women...............................30
Glucocorticoid Induced Osteoporosis......................................34
Osteoporosis in Patients with CKD..40
Treatment: Calcium..44
Treatment: Vitamin D..46
Treatment: Estrogen...48
Treatment: Bisphosphonates..51
Treatment: Denosumab..57
Treatment: Teriparatide..61
Fragility Fracture Overview and Management.........................65
Patient Osteoporosis Questionnaire..71
Patient Handout: Treatment of Osteoporosis..........................72
References..74

Preface

Osteoporosis is a prevalent medical condition characterized by decreased bone mineral density and increased fracture risk. While there is no cure for osteoporosis, proper identification and management of the disease reduces the incidence of fragility fractures and the associated patient morbidity and mortality. The presence of an aging population ensures that medical providers in multiple medical specialties will be required to identify and manage patients with osteoporosis. This text provides medical professionals with the latest treatment guidelines from respected medical organizations. The information is presented in a format that allows for rapid reference and identification of critical information. This text includes sections dedicated to the assessment of fracture risk, secondary causes of osteoporosis, as well as the diagnosis, treatment, and long term management of osteoporosis in multiple patient populations including men, premenopausal women, postmenopausal women, patients with chronic kidney disease and those taking glucocorticoids. Additional text detailing the diagnosis and treatment of fragility fractures, including those involving the distal radius, proximal humerus, hip, pelvis, and vertebrae is provided. A sample patient questionnaire that may be used in clinical practice to identify risk factors for secondary osteoporosis is included, along with an educational handout for patients regarding the nonpharmaceutical treatment of osteoporosis. Following completion of this text, the clinician will be armed with the necessary information to manage the complexities of osteoporosis care.

1 General Information

What is Osteoporosis?
- Osteoporosis is a medical condition characterized by a reduction in bone mass and deterioration of the tissue comprising the skeletal system[1]
- Bone mineralization remains normal, however, there is a reduction in the quantity of bone
- There is no cure for osteoporosis → the therapeutic focus is on decreasing fracture risk and the associated patient morbidity/mortality

What is a Fragility Fracture?
- A fragility fracture is a fracture that occurs from a physiologic stress that one would not expect to lead to a fracture
- This typically involves a fall from a standing height or less leading to a fracture of the proximal humerus, distal radius, proximal femur, pelvis, or vertebrae
- Fragility fractures are often the initial presentation of osteoporosis, which should prompt the astute clinician to consider further diagnostic testing

Epidemiology
- Individuals reach peak bone mass at approximately 30 years of age, which is followed by a progressive decrease in bone mass and increase in fracture risk as one ages[2]
- Over 10 million people in the US and 200 million people worldwide are estimated to currently suffer from osteoporosis[3,1]
- 30-40% of postmenopausal women, and 15-30% of men >50 years old will experience an osteoporosis related fracture[4,5]
- Over 1.5 million people in the US and 8.9 million people worldwide suffer fragility fractures yearly, which results in an osteoporotic fracture every 3 seconds[3,6]
- By 2050, the worldwide incidence of hip fracture is projected to increase by 310% and 240% in men and women, respectively[7]
- With each standard deviation decrease in bone mineral density (BMD), the fracture risk increases 100%[8]
- Following a fragility fracture, the risk of subsequent fractures increases up to 86%[9]

Classification of Osteoporosis (Riggs and Melton)[10]
- ***Type I Primary Osteoporosis:*** Postmenopausal osteoporosis
    - More metabolically active trabecular bone is lost > cortical bone
    - Increased risk of spine and distal radius fractures compared to hip fractures as the femoral neck contains more cortical bone than the spine/distal radius
- ***Type II Primary Osteoporosis***: Elderly >70 years old
    - Affects both trabecular and cortical bone
    - In women, the number of trabeculae decrease whereas in men, the trabeculae are thinned[11]
    - Increased risk of spine, distal radius, and hip fractures
- ***Secondary Osteoporosis***
    - Overall, secondary osteoporosis accounts for approximately 40% of fragility fractures[12]
    - Up to 75% of men and 30-50% of women have a secondary cause for osteoporosis[13-15]
    - A variety of secondary causes of osteoporosis exist, which requires careful consideration when evaluating a patient with osteoporosis and/or fragility fractures

Anatomy/Physiology
- Bone is an active organ system that undergoes a constant remodeling process involving the resorption of old/damaged bone and ossification of new bone matrix
- Approximately 99% of the body's calcium stores reside in bone tissue
- **Cortical bone**: dense outer shell of bone
- **Trabecular/Cancellous bone**: less dense inner lattice-like structure of bone found at the ends of long bones and pelvis
- **Osteoblasts**: bone cells that synthesize new bone tissue
- **Osteoclasts**: bone cells that resorb mature/damaged bone tissue
- **Bone resorption**: process by which osteoclasts break down bone and release the mineral content from the bone into the bloodstream
- **Bone ossification**: process by which osteoblasts lay down new bone material
- **RANKL-RANK-OPG Pathway**[16]
    - "Communication" between osteoblasts and osteoclasts aids in controlling the rate of bone resorption
    - RANKL is secreted by osteoblasts and binds to RANK on osteoclast precursor cells

- RANKL+RANK→ Simulates differentiation of osteoclasts, **activates bone resorption**
- OPG is secreted by osteoblasts and binds to RANKL
  - RANKL+OPG→ Inhibits RANKL binding to RANK→ **decreases bone resorption**

2 Screening for Osteoporosis

Overview
- Screening for osteoporosis begins with identification of at risk populations for osteoporosis and fragility fractures
- A general guideline for identifying which patient populations to screen is presented below. Refer to the sections on the treatment of specific patient populations for detailed recommendations.
- Following identification of high risk patients, appropriate screening consists of a thorough patient history, laboratory analysis, and DXA imaging
- The patient history should identify modifiable and nonmodifiable risk factors associated with the development of osteoporosis and fragility fractures, as well as risk factors for secondary osteoporosis
- The laboratory analysis should ensure adequate serum calcium and vitamin D while screening for secondary causes of osteoporosis
- DXA imaging provides an objective measure of bone density which can be used to categorize patients into fracture risk stratifications, as well as monitor response to treatment

Screening for Osteoporosis: Patient History
- *Who to Screen?*
    - All women >65 years old
    - Women <65 years old with risk factors for fracture
    - Men with risk factors for fracture

- *Modifiable Risk Factors for Osteoporosis[17]*
    - **Alcohol consumption**: >2 drinks daily leads to a 40% increased risk of fragility fracture[18]
    - **Smoking**: 1 PPD in adult life leads to a 5-10% loss of BMD and increased risk of fracture[19,20]
    - **Low body mass index**: BMI <20
    - **Nutrition**: poor calcium intake, vitamin D deficiency
    - **Insufficient exercise**: weightbearing exercise improves BMD, muscular strength, and decrease falls
    - **Frequent falls**

- *Nonmodifiable Risk Factors for Osteoporosis[17]*
    - **Genetics:** family history of osteoporosis

- **Ethnicity**: osteoporosis is more common in Caucasian and Asian populations
- **Age**
- **Gender**
- **Previous fracture**
- **Menopause**
- **Rheumatoid arthritis**

Screening for Osteoporosis: Imaging Modalities
- *Dual-energy x-ray absorptiometry (DXA)*[21]
  - DXA is an imaging modality which uses ionizing radiation (x-ray) to measure bone areal density (g/cm^2)
  - DXA scans are routinely performed at two axial locations (lumbar spine and hip)
  - In select patient populations, DXA scans of the distal third radius can be considered
  - Hip measurements are more predictive of hip and overall osteoporotic fracture risk
  - Spine measurements are more useful in monitoring treatment response, as vertebrae have more metabolically active trabecular bone compared to the hip, leading to an earlier and more robust increase in BMD at this location
  - Diagnosis of osteoporosis is based on the lowest obtained T-score from the lumbar spine, hip, or distal third radius
  - T-scores from locations other than the femoral neck, total femur, lumbar spine, or distal third radius cannot be used for the diagnostic classification of osteoporosis

- *DXA T-score Interpretation*
  - A T-score is a comparison of a patient's BMD in relation to the healthy young adult peak BMD mean
  - The scores are reported in standard deviations from the mean
  - A score of 0 represents average BMD compared to the healthy young adult peak BMD mean
  - A negative score represents a BMD less than the average when compared to the healthy young adult peak BMD
  - A positive score represents a BMD greater than the average when compared to the healthy young adult peak BMD
  - Clinical use: A T-score <-2.5 in **postmenopausal women and men >50 years old** confirms the diagnosis of

osteoporosis. A T-score of -1.0 to -2.5 indicates low bone mineral density (osteopenia). Normal BMD is defined as a T-score greater than -0.9.

- *DXA Z-score Interpretation*
    - A Z-score is a comparison of a patient's BMD in relation to an aged matched BMD mean
    - The scores are reported in standard deviations from the mean
    - A score of 0 represents average BMD when compared to the age-matched average BMD
    - A negative score represents a BMD less than the age-matched average BMD
    - A positive score represents a BMD greater than the age-matched average BMD
    - Clinical use: A z-score of <-2.0 in **premenopausal women and men <50 years old** should raise suspicion for osteoporosis. A z-score <-2.0 in any patient should raise suspicion for secondary causes of osteoporosis.

- *Recommendations for Follow up DXA Testing* [22]
    - The recommended interval between DXA screenings is dependent upon the initial DXA T-score and clinical risk factors for ongoing bone loss and fragility fracture

T-score -2.0 to -2.49 or risk factors for ongoing bone loss
• Follow up DXA scan every 2 years
• Average between baseline testing and development of osteoporosis: 1.1 years
T-score -1.50 to -1.99, no risk factors for accelerated bone loss
• Follow up DXA scan in 3-5 years
• Average between baseline testing and development of osteoporosis: 4.7 years
T-score -1.01 to -1.49, no risk factors for accelerated bone loss
• Follow up DXA scan in 10-15 years
• Average between baseline testing and development of osteoporosis: 17 years

- *Quantitative CT[21]*
    - Quantitative computerized tomography (QCT) is an imaging modality which uses ionizing radiation to measure BMD
    - Quantitative CT is generally only used in research settings given the high cost and significant radiation exposure
    - Spine BMD measured by QCT has the same predictive value for vertebral fractures as a lumbar DXA in postmenopausal women. There is a lack of sufficient evidence for the predictive value in men.
    - There is a lack of sufficient evidence to recommend spine QCT for hip fracture prediction in women or men
    - QCT determined BMD of the lumbar spine can be used to monitor changes in BMD over time

- *Quantitative US[21]*
    - Quantitative ultrasound (QUS) is an imaging modality which uses high frequency ultrasound waves to measure BMD
    - QUS lacks the sensitivity and specificity as other imaging modalities, but it is convenient and does not involve the use of ionizing radiation
    - It is not uncommon for discordant results between QUS and DXA in the same individual
    - Validated heel QUS devices predict fragility fracture in postmenopausal women (hip, vertebral, and global fracture risk) and men over the age of 65 independently of DXA-determined BMD
    - QUS cannot be used to monitor the skeletal effects of treatments for osteoporosis

Screening for Osteoporosis: Laboratory Analysis
- The laboratory analysis should ensure adequate serum calcium and vitamin D while screening for secondary causes of osteoporosis

Routine Laboratory Screening in Osteoporosis	
Mandatory Labs	<ul><li>**CBC**: Evaluate for unexplained anemia</li><li>**CMP**: Evaluate renal and hepatic function, serum calcium and phosphorus.</li><li>**25-hydroxyvitamin D**: Evaluate for vitamin D deficiency</li><li>**PTH**: Evaluate for hyperparathyroidism</li></ul>

	- **TSH**: Evaluate for hyperthyroidism
- **24-hour urine calcium**: Evaluate for idiopathic hypercalciuria
Lab Considerations	- **Fasting morning testosterone, LH/FSH**: Evaluate male hypogonadism
- **Tissue transglutaminase antibody (tTG)**: Evaluate for celiac sprue
- **Serum/urine protein electrophoresis**: Evaluate for multiple myeloma
- **Salivary cortisol**: Evaluate for Cushing's syndrome

- *Biochemical Bone Turnover Markers (BTMs)*
    - BTMs are biochemical by-products of bone remodeling released into systemic circulation following bone formation or resorption
    - BTMs are useful for monitoring response to antiresorptive and anabolic treatment, particularly in patients with malabsorption, and in patients in which compliance is a concern
    - BTMs are not approved for establishing the diagnosis of osteoporosis
    - Measurement of BTMs can confirm response to treatment within 3-6 months, whereas a stable or improving BMD on DXA following 2 years of therapy confirms adequate treatment response
    - BTMs demonstrate significant biologic and laboratory variability, which should be considered when interpreting the results

Bone Turnover Markers in Osteoporosis	

BTMs Specific to Bone Resorption[23,24]	• **Urinary NTX: N-terminal telopeptide of type I collagen**
	    o Response to treatment at 3-6 months: -50% compared to baseline
	• **Serum CTX: C-terminal telopeptide of type I collagen**
	    o Response to treatment at 3-6 months: -30% compared to baseline

BTMs Specific to Bone Formation[24,25]	• **Serum P1NP: type 1 procollagen (N-terminal)** o Response to treatment at 3-6 months: + ≥21% or increase >10 mcg/L compared to baseline

3 Assessment of Fracture Risk

Overview
- In patients with low BMD, it is imperative to determine the risk of fracture as this will help guide treatment recommendations and follow up care
- Fracture risk is multivariate, including numerous contributing factors which decrease bone density and increase fall risk
- **Proximal femur morphology and fracture risk**[26]
    - Hip axis length: each standard deviation above the mean of control population increases risk of femoral neck fracture 1.9-fold and intertrochanteric fracture 1.6-fold
    - Femoral neck-shaft angle and femoral neck width may also affect fracture risk
- **Advancing age and fracture risk**[27]
    - Age affects fracture risk irrespective of BMD
    - Older patients with same T-score as a younger population will have higher fracture risk
    - An 80 year old patient with a T-score of -2.5 has a fracture risk 5-fold greater than a 50 year old patient with the same T-score
- The fracture risk assessment tool (FRAX) and the Garvan fracture risk calculator (FRC) are useful clinical tools to accurately determine fracture risk and aid in treatment decisions

	Garvan
Function	- Predicts 5 + 10 year probability of sustaining fragility fractures[28]
Risk Factors	- Age, weight, fracture after age 50, falls in last 12 months, BMD (T-score hip or lumbar spine)
Population	- Sex: male and female - Age: 60+ years old
Strengths/Limitations	- Allows for T-score from spine or hip - Quantifies falls, fragility fractures - Fewer risk factor inputs has questionable effect on result accuracy
Website	- http://www.garvan.org.au/promotions/bone-fracture-risk/calculator/

FRAX (Gold Standard)	
Function	• Estimates 10 year risk of hip and major osteoporotic fracture
Risk Factors	• Age, weight, height, parental hip fracture, personal fragility fracture, cigarette smoking, alcohol abuse, RA, corticosteroid use, secondary osteoporosis
Population	• Sex: male and female • Age: 40-90 years old
Strengths/Limitations	• Results are only accurate in patients who have not previously taken osteoporosis medication • Rigid yes/no answering input without ability to quantify • Only uses femoral neck T-score
Website	• http://www.shef.ac.uk/FRAX/

4 Secondary Osteoporosis: Medical Conditions

Overview
- Secondary osteoporosis is defined as low BMD caused by an underlying medical condition or use of medication which is known to affect BMD
- In men, secondary causes of osteoporosis are present in up to 75% of patients[29]
- In women, secondary causes of osteoporosis are present in up to 30-50% of patients[30,31]
- A thorough history, physical examination, and laboratory analysis is required to identify secondary causes of osteoporosis
- Z-scores (comparison of a patient's BMD in relation to an aged matched BMD mean) of <-2.0 in any patient should raise suspicion for secondary causes of osteoporosis

Nonpharmaceutical Causes of Secondary Osteoporosis
• **Cushing's syndrome** • **Hyperthyroidism** • **Primary hyperparathyroidism** • **Malabsorption—celiac disease** • **Hypercalciuria** • **Diabetes mellitus: type I and type II** • **Multiple myeloma** • **Mastocytosis** • Family history of low trauma fractures • Excessive alcohol consumption (>2 drinks daily) • Smoking • Low calcium/Vitamin D intake • Primary/secondary hypogonadism • Low BMI (<20), associated eating disorders • Lack of or excessive exercise • Chronic liver or kidney disease • Rheumatoid arthritis • HIV, treatment with protease inhibitors • Organ transplant • Osteogenesis imperfecta

Diabetes Mellitus (DM)
- Type 1 DM is an autoimmune destruction of beta islet cells in the pancreas which leads to inadequate production of insulin
- Type 2 DM is an inadequate response of end organ systems at cellular level to insulin
- Symptoms
    - Increased thirst, increased hunger, frequent urination, unexplained weight loss
- Diagnosis
    - HgbA1C ≥6.5, fasting glucose ≥126 mg/dL
- **Osteoporosis Considerations**
    - Type 1 patients have 12x greater risk of sustaining osteoporotic fractures[32]
    - Type 2 patients have a 20% increase in fracture risk[32]
    - DM patients have an increased fall risk from neuropathy, retinopathy

Hyperthyroidism
- Hyperthyroidism is excessive thyroid gland production of thyroid hormone
- Symptoms
    - Skin: onycholysis (loosening of nail from nailbed), hyperpigmentation, thinning of hair, infiltrative dermopathy (pretibial edema, hyperpigmented, orange-peel textured papules)
    - Eyes: lid lag, exophthalmos
    - Cardiovascular: tachycardia, systolic hypertension
    - Anxiety, restlessness, irritability, insomnia, goiter, weight loss
- Diagnosis:
    - **TSH** : low (< 0.4 mIU/L), **Free T4**: high (>1.8 ng/dL)
- **Osteoporosis Considerations**
    - Activation of thyroid hormone receptor alpha on osteoblasts and osteoclasts increases bone resorption[33]
    - Serum TSH <0.1 mU/L is associated with a 4-fold greater risk of hip fracture, 5-fold greater risk of spine fracture[34]

Primary Hyperparathyroidism
- Primary hyperparathyroidism is excessive parathyroid gland production of parathyroid hormone (PTH)
- Secondary hyperparathyroidism involves elevated PTH levels secondary to CKD or vitamin D deficiency

- Symptoms
    - Most common clinical presentation: **asymptomatic hypercalcemia**
        - If asymptomatic, serum calcium is generally <1.0 mg/dL above normal limits
    - Classic symptoms including "bones, stones, groans, psychiatric moans" are now uncommon in developed countries
    - Nonspecific symptoms including fatigue, weakness, anorexia, mild depression, mild cognitive/neuromuscular dysfunction have been reported[35]
    - **Osteitis fibrosa cystica**
        - Develops with longstanding, marked elevations in PTH
        - Symptoms include bone pain, subperiosteal bone resorption on radial aspect of middle phalanx
- Diagnosis
    - Primary Hyperparathyroidism
        - **Serum PTH**: elevated (> 65 pg/mL)
        - **Serum calcium (albumin corrected)**: elevated (>10.2 mg/dL)
        - Phosphate: low normal – PTH inhibits proximal tubular reabsorption of phosphate
        - 25-hydroxyvitamin D: upper normal or elevated
        - Magnesium: low normal – renal tubular reabsorption of magnesium inhibited by hypercalcemia
    - Secondary Hyperparathyroidism
        - **Serum PTH**: elevated
        - **Serum calcium (albumin corrected)**: low/low normal
        - If CKD: elevated creatinine, decreased eGFR
        - If vitamin D deficiency: low 25-hydroxyvitamin D
- **Osteoporosis Considerations**
    - Women are affected by primary hyperparathyroidism at a 4-fold greater incidence than men[36]
    - Chronic, continuous PTH excess increases bone resorption through stimulation of osteoclast-activating factors from osteoblasts[37]
    - Cortical bone is affected to a greater extent than cancellous bone[37]

- Normocalcemic hyperparathyroidism has been reported and appears to carry a comparable level of fragility fracture risk as that observed in primary hyperparathyroidism[38]

Idiopathic Hypercalciuria
- Idiopathic hypercalciuria is excessive urinary calcium excretion without an identifiable underlying cause
- Symptoms
    - Generally asymptomatic
    - Increased incidence of renal stones
- Diagnosis
    - Diagnosis of exclusion: must confirm no underlying cause (renal disease, hyperparathyroidism, etc.)
    - **24-hour urinary calcium excretion**: >4mg/kg/day
    - Serum PTH: normal
    - Serum calcium (albumin corrected): normal
    - Creatinine/eGFR: normal
- **Osteoporosis Considerations**
    - Patients with primary hypercalciuria have been shown to have decreased BMD and increased fracture risk[39]
    - Use of Thiazide diuretics may reduce urinary calcium excretion and improve BMD[40]

Primary/Secondary Male Hypogonadism
- Primary hypogonadism is a reduction in testicular production of testosterone
- Secondary hypogonadism is a reduction in testosterone from inadequate pituitary or hypothalamus production of LH/FSH
- Symptoms
    - **Decreased libido**, infertility, decreased testicular size, decreased energy, decreased body hair, decreased muscle mass, gynecomastia
- Diagnosis
    - Primary hypogonadism
        - **Testosterone**: low
        - **LH/FSH**: elevated
    - Secondary hypogonadism
        - **Testosterone**: low
        - **LH/FSH**: normal/low

- **Osteoporosis Considerations**
    - Male hypogonadism is a major risk factor for low BMD and fragility fractures in men as low serum testosterone affects peak bone mass accrual and maintenance of bone strength[41]

Celiac Disease
- Celiac disease is an autoimmune disorder in which antibodies are formed against gluten leading to inflammation and villous atrophy in the small intestine
- Symptoms
    - Patients may be asymptomatic
    - Chronic diarrhea/steatorrhea
    - Unexplained weight loss
- Diagnosis
    - **Tissue transglutaminase antibody (tTG): Positive**
    - Consider: Quantitative IgA, anti-DGP, EMA, ARA
- **Osteoporosis Considerations**
    - Prevalence of celiac disease is 3.4% in patients with osteoporosis vs. 0.2% in patients without osteoporosis[42]
    - Patients may experience chronic malabsorption of calcium and vitamin D

Myeloproliferative Disease: Multiple Myeloma + MGUS
- Multiple myeloma is a malignant proliferation of plasma cells in bone marrow with production of a monoclonal immunoglobulin
- Symptoms
    - Initially, patients may be asymptomatic
    - Patients may present with bone pain and fractures, often involving the axial skeleton (spine, ribs)
- Diagnosis
    - **Serum protein electrophoresis**
    - **Urine protein electrophoresis**
    - CBC: unexplained anemia
    - Bone marrow biopsy
- **Osteoporosis Considerations**
    - Expression of pro-osteoclastogenic factors and cytokine secretion by myeloma cells increase bone resorption[43]
    - Secretion of dickkopf 1 suppresses osteoblast differentiation[44]

Cushing's Syndrome
- Cushing's syndrome is caused by excessive production of cortisol by the adrenal glands
- Symptoms
    - Central adiposity, with fatty deposits around the face (moon face) and upper back (hump back)
    - Muscular weakness and easy fatigability
    - Thin skin with easy bruising, stretch marks, delayed healing
    - Women may experience hirsutism, irregular menstruation
    - Hypertension, diabetes, hyperlipidemia
- Diagnosis
    - **Salivary cortisol (11 pm)**: elevated
    - 24-hour urinary free cortisol: elevated
    - Overnight dexamethasone suppression: elevated
- **Osteoporosis Considerations**
    - Chronic exposure to elevated levels of glucocorticoids leads to both increased bone resorption and decreased bone formation which decreases BMD and increases fracture risk (see section Glucocorticoid Induced Osteoporosis)

Systemic Mastocytosis
- Systemic mastocytosis is the abnormal proliferation of mast cells in multiple body tissues
- Mast cells release chemical mediators, including histamine, during allergic reactions
- Symptoms
    - Most symptoms are nonspecific, including abdominal pain, GERD, pruritus, anaphylactic reaction
    - May present with bone pain, unexplained osteoporosis
- Diagnosis
    - **Serum Tryptase**: elevated
- **Osteoporosis Considerations**
    - Approximately 50% of patients with systemic mastocytosis will have bone manifestations[45]
    - Osteoporosis develops secondary to excessive degranulation of mast cell products (interleukins, histamine) which increase osteoclast differentiation and bone resorption[46]

5 Secondary Osteoporosis: Medications

Overview
- While glucocorticoids have a well established reputation of reducing BMD and increasing fracture risk, a number of other commonly prescribed medications must also be considered
- Medications vary in the severity of their effect on BMD and each requires careful consideration when evaluating fracture risk and determining appropriate therapeutic options

Pharmacologic Causes of Secondary Osteoporosis
- Aromatase Inhibitors: Anastrazole, Letrozole, Exemestrane
- Antidepressants: Sertraline, Citalopram, Paroxetine, Fluoxetine
- Anti-Epileptic Drugs; Carbamazepine, Valproate, Phenobarbital, Phenytoin
- Calcineurin Inhibitors: Cyclosporine, Tacrolimus
- Depot Medroxyprogesterone Acetate
- Glucocorticoids: Prednisone, Methylprednisolone
- GnRH Agonists: Leuprorelin, Triptorelin, Goserelin,
- Proton Pump Inhibitors: Omeprazole, Pantoprazole
- Thiazolidinediones: Rosiglitazone

Aromatase Inhibitors (AI)
- Pharmacologic use: therapy for estrogen receptor positive breast cancer, ovarian cancer, gynecomastia
- Mechanism: inhibits the function of the enzyme aromatase, which converts androgens to estrogens (net result: decreased serum estrogen)
- **Osteoporosis Considerations**
    - Decreased levels of estrogen accelerates bone turnover and is associated with increased fracture risk[47,48]
    - The current recommendation for women on AIs is assessment of fracture risk in all patients, and use of osteoporosis medication in all patients with a T-score <-2.5 or if FRAX risk meets treatment criteria, as well as all women >75 years old[49]

Antidepressants (SSRIs)
- Pharmacologic use: therapy for depression and anxiety disorders
- Mechanism: selectively inhibits the reuptake of serotonin
- **Osteoporosis Considerations**
    - SSRIs negatively impact osteoblast formation and function[50]
    - Rate of bone loss is 1.6-fold greater in patients taking SSRIs when compared to a control population[51]

Anti-Epileptic Drugs (AEDs)
- Pharmacologic use: therapy for seizure disorders, bipolar disorder
- Mechanism: variable dependent upon medication
- **Osteoporosis Considerations**
    - AEDs are associated with increased bone turnover, decreased BMD, and increased fracture risk[52,53]
    - Enzyme inducing AEDs (carbamazepine, valproate, phenobarbital, phenytoin) are associated with a 2-6-fold increase in fracture risk[54]

Calcineurin Inhibitors (CIs)
- Pharmacologic use: therapy for transplant patients
- Mechanism: immunomodulator→ inhibits T-lymphocyte activation
- **Osteoporosis Considerations**
    - CIs increase bone resorption and decrease BMD[55]
    - One study demonstrated that nearly 60% of patients experienced a fracture after solid organ transplant within 6.5 years of starting glucocorticoid-sparing immunosuppressive therapy which included CIs[56]

Depot Medroxyprogesterone Acetate (DMPA)
- Pharmacologic use: intra-muscular hormonal contraceptive
- Mechanism: inhibits the secretion of pituitary gonadotropin→ prevents follicular maturation and ovulation, thickens cervical mucus
- **Osteoporosis Considerations**
    - Low serum gonadotropin levels→ increases bone turnover and decreases BMD[57]
    - BMD levels return to baseline following cessation of therapy[57]

- o Patients with greatest fracture risk are women >30yo with >10 prior DMPA injections[58]

GnRH Agonists
- Pharmacologic use: prostate cancer, endometriosis
- Mechanism: suppression of gonadotropin secretion
- **Osteoporosis Considerations**
  - o Decrease serum gonadotropin levels increases bone turnover and decreases BMD
  - o Men treated with GnRH agonists for prostate cancer may have up to a 60% increased fracture risk[59]

Proton Pump Inhibitors (PPI)
- Pharmacologic use: therapy for GERD, PUD, gastritis
- Mechanism: inhibits gastric parietal cell hydrogen-potassium ATPase
- **Osteoporosis Considerations**
  - o Acid suppression may impair calcium absorption
  - o Modest association between PPI use and increased risk of hip and vertebral fractures[60]

Thiazolidinediones (TZD)
- Pharmacologic use: therapy for type 2 DM
- Mechanism: increases insulin sensitivity
- **Osteoporosis Considerations**
  - o Decreases bone formation by decreasing osteoblast function[61]
  - o Studies have reported an increased fracture risk in women taking TZDs[62]

6 Osteoporosis in Men

Background Information
- More than 8 million men in US have low bone mineral density or osteoporosis[63]
- Men over the age of 50 carry a 10-25% lifetime risk of fragility fracture[64]
- Men treated with androgen deprivation therapy (ADT) for prostate cancer have a 20% fracture risk within 5 years of treatment initiation[65]
- Men account for 25% of hip fractures[66]
- Men experience increased morbidity and mortality following hip fractures compared to women, with a 2-fold greater risk of death following a hip fracture[67]

Pathogenesis
- Men do not experience menopause and the subsequent abrupt loss of sex hormones (estrogen)
- Instead, under normal physiologic conditions there is a gradual loss of testosterone production over time
- As a result, men acquire a greater peak BMD and a more gradual decline in BMD than women
- There is a correlation between deficient serum testosterone and estradiol and fracture risk in men[68]
- Studies indicate that up to 75% of men have a secondary cause of osteoporosis, which requires a thorough history, examination, and appropriate imaging/laboratory analysis[69]

Clinical Manifestations
- Typically no symptoms until fracture
- Consider manifestations of secondary causes of osteoporosis

Osteoporosis Screening in Men

	DXA
Routine Screening	• Routine screening is not recommended, even in patients >70years old[70]
Indications for DXA	• **Fragility fracture** (hip, pelvis, spine, proximal humerus, distal radius)

	- **Glucocorticoid treatment** (≥5mg for ≥3 months)
- **Androgen deprivation therapy** (ADT) for prostate cancer
- **Include distal third radius BMD in men on ADT**[71]
- Radiographic osteopenia
- Height loss >1.5 inches from maximum height
- Symptomatic hypogonadism: decreased libido
- Current smoking/COPD
- Alcoholism
- Malabsorption or bariatric surgery
- Use of medications associated with secondary osteoporosis
- Recurrent renal stones/hypercalciuria
DXA Interpretation	- Men have larger bones, and may have osteophytes, facet sclerosis, DISH syndrome which may raise apparent BMD and underestimate fracture risk[72]

Screening for Osteoporosis in Men	
Imaging	- DXA: hip, lumbar spine +/- distal third radius (men on ADT)
Mandatory Labs	- CBC
- CMP: Creat/GFR, alkaline phosphatase, phosphate, calcium (with serum albumin for calculation of corrected serum calcium)
- 25-hydroxyvitamin D
- PTH
- TSH
- 24-hour urine calcium
Lab Considerations	- +/- Fasting morning testosterone, LH/FSH (symptoms of hypogonadism)
- +/- Tissue transglutaminase antibody (tTG) (symptoms of malabsorption/celiac)
- +/- Serum/urine protein electrophoresis (symptoms of multiple myeloma)
- +/- Salivary cortisol (symptoms of Cushing's syndrome)

	• +/- Serum P1NP/CTX, urinary NTX to monitor response to therapy

Diagnostic Classification in Men
- Osteoporosis
    - Men >50 years old: T-score <-2.5[73]
    - Men <50 years old: Z-score <-2.0 + fragility fracture[74]
- Low Bone Mineral Density (Osteopenia)
    - T-score -1.0 to -2.5
    - Must calculate FRAX to determine fracture risk
- Normal BMD
    - T-score 0 to -0.9

Candidates for Pharmacologic Therapy[75]
- T-score <-2.5
- T-score -1.0 to -2.5: FRAX 10-year hip fracture risk >3% or 10 year major osteoporotic fracture risk >20%
- History of fragility fracture
- Men receiving long term glucocorticoid therapy (≥7.5 mg daily)

Male Osteoporosis Treatment Protocol[75]
Lifestyle Modifications
1) Weightbearing exercise program
2) Smoking cessation
3) Modest alcohol intake (<2 drinks daily)
4) Calcium: **1000-1200** mg daily
5) Vitamin D: **600-800** IU daily
6) Osteoporosis specific medication (see chart below)

Osteoporosis Medication
First Line
• **Alendronate** (70mg po weekly) or **Risedronate** (35mg po weekly)
Second Line
• **Zoledronic acid** (5mg IV yearly)→ if contraindication to oral bisphosphonates
Third Line
• **Denosumab** (60mg SC q6mo→ only current indication is men treated with ADT for prostate cancer
• **Teriparatide** (20mcg SC daily)→ consider as first line therapy in severe osteoporosis with prior fragility fracture

Hormone Replacement Therapy (HRT)
- **Testosterone** (50-400mg IM q2-4 weeks)
- Consider only in younger men with clinical manifestations of hypogonadism (decreased libido)
- High risk hypogonadal men on HRT→ consider addition of osteoporosis medication in following:
- High dose glucocorticoids (≥5mg/day x ≥3 months)
- Frequent falls
- Recent fragility fracture
- T-score <-2.5
- Persistent T-score <-2.5 despite 2 years of HRT

Monitoring Response to Therapy
Prior to Initiation of Medication
- Obtain baseline DXA: 2 anatomic sites (hip, lumbar spine, +/- distal third radius)
- Consider measuring bone turnover markers (NTX, CTX) if starting antiresorptive medication
- Strongly consider in patients with malabsorption or questionable compliance
- Obtain prior to initiation of treatment, and repeat at 3-6 months
- Adequate response to treatment[76]
- NTX: >50% decrease compared to baseline
- CTX: >30% decrease compared to baseline
After Initiation of Medication
- Obtain DXA at 2 year intervals following initiation of medication
- 2 anatomic sites (hip, lumbar spine +/- distal third radius)
- Preference is to obtain follow up DXA using same DXA scanner for accurate comparison to prior imaging
Drug Holiday (Bisphosphonates)
- Given the long half-life of bisphosphonates, it is reasonable to consider a drug holiday for select patients
- Criteria: use of **Alendronate/Risedronate** x 5 years, or **Zoledronic acid** x 3 years with a stable/improving BMD, no fragility fractures, and low risk for fracture in near future
- Obtain DXA 2 years following cessation of therapy
- Restart bisphosphonates with persistent bone loss (>5% on 2 subsequent DXAs 2 years apart) or if the patient sustains a fragility fracture during the drug holiday

7 Osteoporosis in Premenopausal Women

Background Information
- The average age of female menopause is 48-55 years old[77]
- The peak period of bone mass accrual is during the teenage years and the majority of bone mass accrual has taken place by 20 years of age[78]
- Women with a premenopausal fracture are 35% more likely to sustain postmenopausal fractures[79]

Pathogenesis
- Premenopausal osteoporosis can be secondary to **inadequate peak bone mass acquisition** or from **ongoing bone loss** which requires consideration of secondary causes of osteoporosis
- It is normal for women to have a loss of BMD during pregnancy (3-5%) and lactation (3-10%), however, this is reversed following weaning and no known link between the number of pregnancies and osteoporosis risk has been established[80]
- Pregnancy and lactation associated osteoporosis (PLO) and transient osteoporosis of the hip are rare clinical conditions which can present with hip pain or fragility fracture in the third trimester. These forms of osteoporosis generally resolve spontaneously, but may require short term osteoporosis directed treatment along with management of acute pain and/or trauma[81]

Clinical Manifestation
- Typically no symptoms until fracture
- Consider manifestations of secondary causes of osteoporosis

Osteoporosis Screening in Premenopausal Women

	DXA
Routine Screening	• Routine screening is not recommended for women <65 years old[82]
Indications for DXA	• **Fragility fracture** (hip, pelvis, spine, proximal humerus, distal radius) • **Glucocorticoid treatment** (≥5mg for ≥3

	months) • **Chemotherapy:** cyclophosphamide, methotrexate, 5-fluorouracil, GnRH agonists • Radiographic osteopenia • Medical conditions associated with secondary osteoporosis • Use of medications associated with secondary osteoporosis
DXA Interpretation	• Women <50 years old should use Z-score for interpretation of osteoporosis

Screening for Osteoporosis in Premenopausal Women	
Imaging	• DXA: hip, lumbar spine (+/- distal third radius)
Mandatory Labs	• CBC • CMP: Creat/GFR, alkaline phosphatase, phosphate, calcium (with serum albumin for calculation of corrected serum calcium) • 25-hydroxyvitamin D • PTH • TSH • 24-hour urine calcium
Lab Considerations	• +/- LH/FSH, estrogen (if amenorrhea) • +/- Tissue transglutaminase antibody (tTG) (symptoms of malabsorption/celiac) • +/- Serum/urine protein electrophoresis (symptoms of multiple myeloma) • +/- Salivary cortisol (symptoms of Cushing's syndrome) • +/- Serum P1NP/CTX or urinary NTX to monitor response to therapy

Diagnostic Classification in Premenopausal Women[83]
- Relationship between BMD and fracture risk is not the same as that of postmenopausal women
- Premenopausal Osteoporosis
  - Z-score <-2.0 + fragility fracture

Candidates for Pharmacologic Therapy
1) Z-score <-2.0 and one of the following:
   o Fragility fracture
   o Active bone loss
   o Ongoing secondary cause of osteoporosis (medical condition/medication)

Premenopausal Osteoporosis Treatment Protocol
Lifestyle Modifications
1) Weightbearing exercise program
2) Smoking cessation
3) Modest alcohol intake (<2 drinks daily)
4) Calcium: **1000-1200** mg daily
5) Vitamin D: **600-800** IU daily
6) Osteoporosis specific medication (see chart below)

Osteoporosis Medication: Premenopausal Women

Amenorrhea in Younger Premenopausal Women
• **Estrogen only vs. Estrogen-Progestin combination** (Dosing and route varies)
Glucocorticoids (≥7.5mg daily, +/- fragility fracture)
• **Alendronate** (70mg po weekly) or **Risedronate** (35mg po weekly) • **Zoledronic acid** (5mg IV yearly)→ if contraindication to oral bisphosphonates • **Teriparatide** (20mcg SC daily)→ severe osteoporosis + fracture, unresponsive to other therapies • +/- **Estrogen-Progestin combination** (dosing and route varies)→ if associated amenorrhea • See section Glucocorticoid Induced Osteoporosis for specific recommendations
Chemotherapy
• **Alendronate** (70mg po weekly) or **Risedronate** (35mg po weekly) • **Zoledronic acid** (5mg IV yearly)→ if contraindication to oral bisphosphonates

Monitoring Response to Therapy

Prior to Initiation of Medication

- Obtain baseline DXA: 2 anatomic sites (hip, lumbar spine, +/- distal third radius)
- Consider measuring bone turnover markers (NTX, CTX) if starting antiresorptive medication
    - Strongly consider in patients with malabsorption or questionable compliance
    - Obtain prior to initiation of treatment, and repeat at 3-6 months
    - Adequate response to treatment[76]
        - NTX: >50% decrease compared to baseline
        - CTX: >30% decrease compared to baseline

After Initiation of Medication

- Obtain DXA at 2 year intervals following initiation of medication
- 2 anatomic sites (hip, lumbar spine +/- distal third radius)
- Preference is to obtain follow up DXA using same DXA scanner for accurate comparison to prior imaging

Drug Holiday (Bisphosphonates)

- Given the long half-life of bisphosphonates, it is reasonable to consider a drug holiday for select patients
- Criteria: use of **Alendronate**/Risedronate x 5 years, or **Zoledronic acid** x 3 years with a stable/improving BMD, no fragility fractures, and low risk for fracture in near future
- Obtain DXA 2 years following cessation of therapy
- Restart bisphosphonates with persistent bone loss (>5% on 2 subsequent DXAs 2 years apart) or if the patient sustains a fragility fracture during the drug holiday

8 | Osteoporosis in Postmenopausal Women

Background Information
- Over 20 million women in the US have osteoporosis or low bone mineral density[84]
- 30-40% of postmenopausal women >50 years old will experience an osteoporosis related fracture[85]
- Women who have a vertebral fracture have a 19% chance of another fracture within 1 year
- 15% of women will experience a hip fracture by age 80, and 5-10% of patients will experience a recurrent hip fracture[86,87]
- National Osteoporosis Foundation treatment guidelines will lead to the recommendation of pharmacologic therapy for 72% of women >65years old and 93% of women >75 years old[88]

Pathogenesis
- Osteoporosis in postmenopausal women occurs secondary to estrogen deficiency, which causes an abrupt onset of excessive bone resorption
- The greatest loss of BMD occurs 2-3 years prior to menopause through 3-4 years post-menopause where the anticipated BMD loss is 10.5% at the spine and 5.3% at the femoral neck[89]
- The overall result of menopause is decreased BMD and increased fracture risk

Clinical Manifestations
- Generally, no manifestations until fracture
- Vertebral fracture is the most common clinical manifestation of osteoporosis, though two-thirds of vertebral compression fractures are asymptomatic

Osteoporosis Screening in Postmenopausal Women

Screening for Osteoporosis in Postmenopausal Women	
Imaging	• DXA: hip, lumbar spine (+/- distal third radius) at age 65, sooner if clinical presence of risk factors

Mandatory Labs	<ul><li>CBC</li><li>CMP: Creat/GFR, alkaline phosphatase, phosphate, calcium (with serum albumin for calculation of corrected serum calcium)</li><li>25-hydroxyvitamin D</li><li>PTH</li><li>TSH</li><li>24-hour urine calcium</li></ul>
Lab Considerations	<ul><li>+/- Tissue transglutaminase antibody (tTG) (symptoms of malabsorption/celiac)</li><li>+/- Serum/urine protein electrophoresis (symptoms of multiple myeloma)</li><li>+/- Salivary cortisol (symptoms of Cushing's syndrome)</li><li>+/- Serum P1NP/CTX vs. urinary NTX to monitor response to therapy</li></ul>

Diagnostic Classification in Postmenopausal Women[90]
- Osteoporosis
    - Radiographic: T-score <-2.5
    - Clinical: diagnosis of osteoporosis if history of fragility fracture
- Low Bone Mineral Density (Osteopenia)
    - T-score -1.0 to -2.5
- Normal BMD
    - T-score 0 to -0.9
- Z-score <-2.0, regardless of T-score, is below the expected BMD range for age and requires careful evaluation for secondary osteoporosis

Candidates for Pharmacologic Therapy[90]
1) T-score <-2.5
2) Clinical presence of fragility fracture
3) T-score -1.0 to -2.5→ calculate FRAX score
    - Identification of high risk patients, as indicated by a FRAX 10 year hip fracture risk >3%, or a 10 year major osteoporotic fracture risk >20%

Postmenopausal Osteoporosis Treatment Protocol
Lifestyle Modifications
1) Weightbearing exercise
2) Smoking cessation
3) Modest alcohol consumption (<2 drinks daily)
4) Calcium: Diet + supplementation **1200** mg daily
5) Vitamin D: **800-1000** IU daily
6) Osteoporosis specific medication (see chart below)

Osteoporosis Medication: Postmenopausal Women
First Line
• **Alendronate** (70mg po weekly) or **Risedronate** (35mg po weekly)
Second Line
• **Zoledronic acid** (5mg IV yearly)→ if contraindication to oral bisphosphonates
Third Line
• **Denosumab** (60mg SC q6mo→ Impaired renal function, unresponsive to other therapies • **Teriparatide** (20mcg SC daily)→ severe osteoporosis + fracture, unresponsive to other therapies
Hormone Replacement Therapy

• **Raloxifene** (60mg po daily)
    ○ Decreases breast cancer risk, increases DVT risk
    ○ Consider when there is an independent need for breast cancer prophylaxis

Monitoring Response to Therapy
Prior to Initiation of Medication

• Obtain baseline DXA: 2 anatomic sites (hip, lumbar spine +/- distal third radius)
• Consider measuring bone turnover markers (NTX, CTX) if starting antiresorptive medication
    ○ Strongly consider in patients with malabsorption or questionable compliance
    ○ Obtain prior to initiation of treatment, and repeat at 3-6 months
    ○ Adequate response to treatment[76]
        ▪ NTX: >50% decrease compared to baseline
        ▪ CTX: >30% decrease compared to baseline

After Initiation of Medication
• Obtain DXA at 2 year intervals following initiation of medication • 2 anatomic sites (hip, lumbar spine +/- distal third radius) • Preference is to obtain follow up DXA using same DXA scanner for accurate comparison to prior imaging
Drug Holiday (Bisphosphonates)
• Given the long half-life of bisphosphonates, it is reasonable to consider a drug holiday for select patients • Criteria: use of **Alendronate/Risedronate** x 5 years, or **Zoledronic Acid** x 3 years with a stable/improving BMD, no fragility fractures, and low risk for fracture in near future • Obtain DXA 2 years following cessation of therapy • Restart bisphosphonates with persistent bone loss (>5% on 2 subsequent DXAs 2 years apart) or if the patient sustains a fragility fracture during the drug holiday

9 | Glucocorticoid Induced Osteoporosis (GIO)

Background Information
- Glucocorticoids are used in 1.2% of the general population and in up to 4.6% of postmenopausal women[91,92]
- Fractures in patients taking oral glucocorticoids occur at higher BMDs than those sustained by the general population[94]
- Bone loss is most rapid within the first 3-6 months of glucocorticoid use[94]
- An estimated 30-50% of patients on chronic glucocorticoids will sustain a fracture[95]
- Chronic glucocorticoid use (≥3 months) and higher doses (≥5mg/day) are associated with increased fracture risk[93]
- Daily dose of glucocorticoids poses a greater fracture risk than cumulative dose of glucocorticoids[96]
- Small decreases in BMD and increased risk of fracture are possible with high doses of inhaled corticosteroids or intermittent corticosteroid use[97,98]
- Termination of steroids is associated with decreased fracture risk and partial reversal of BMD reduction[93]

Pathogenesis of Glucocorticoids
- Glucocorticoids negatively impact bone density by **increasing bone resorption** AND **decreasing bone formation** through a variety of mechanisms
- Initially, trabecular bone is impacted > cortical bone due to its higher metabolic activity[99]
  - Cortical bone is affected with prolonged glucocorticoid exposure[99]
- Early following exposure to glucocorticoids there is an increased fracture risk (particularly vertebral)[93]

Biologic Effects of Glucocorticoids on Bone	
Increased Bone Resorption	• Direct inhibition of osteoblast proliferation and differentiation[100] • Increased apoptosis of osteoblasts and osteocytes[100]

	• Decreased serum estrogen and testosterone[101]
Decreased Bone Formation	• Suppression of OPG synthesis • Stimulates production of RANKL[102] • Decreased secretion of androgens/estrogens (GnRH secretion inhibited)[101] • Decreased IGF-1 • Decreased intestinal calcium absorption[103] • Increased renal calcium excretion[104] • Increased expression of factors that inhibit osteoblast maturation[99]

Osteoporosis Screening in Patients Taking Glucocorticoids

<div align="center">**Screening for Osteoporosis in GIO**</div>	
Imaging	• Baseline DXA (hip, lumbar spine, +/- distal third radius) in all patients >30 years old who take glucocorticoids >3months[105] • Consider vertebral XR imaging in patients with low BMD (T score -1.0 to -2.4), even if no prior symptomatic fracture given the high incidence of vertebral compression fractures in patients on glucocorticoids
Mandatory Labs	• CBC • CMP: Creat/GFR, alkaline phosphatase, phosphate, calcium (with serum albumin for calculation of corrected serum calcium) • 25-hydroxyvitamin D • PTH • TSH • 24-hour urine calcium
Lab Considerations	• +/- Tissue transglutaminase antibody (tTG) (symptoms of malabsorption/celiac) • +/- Serum/urine protein electrophoresis (symptoms of multiple myeloma) • +/- Salivary cortisol (symptoms of Cushing's syndrome) • +/- Serum P1NP/CTX vs. urinary NTX to monitor response to therapy

Candidates for Pharmacologic Therapy[106-108]
- Treatment recommended for all patients with osteoporosis
    - Men >50/postmenopausal women with fragility fracture **OR** T score <-2.5
- Treatment recommended for all high risk patients
    - Men >50/postmenopausal women with T score -1.0 to -2.5
    - FRAX 10-year hip fracture risk >3% or major osteoporotic fracture risk >20%
    - Glucocorticoid dosing ≥7.5mg/day x ≥3 months
- Men <50 and premenopausal women
    - Pharmacologic treatment in men/premenopausal women on ≥7.5mg x ≥ 3 months
        - + fragility fracture
        - >4% BMD loss yearly

Prevention of Osteoporosis in Patients Taking Glucocorticoids
Lifestyle Modifications
1) Decrease dose/duration of exogenous glucocorticoids if medically possible
2) Weightbearing exercise program
3) Smoking cessation
4) Modest alcohol intake (<2 drinks daily)
5) Calcium: **1200** mg daily
6) Vitamin D: **800** IU daily
7) Fall prevention
8) Osteoporosis specific medication (see chart below)

Osteoporosis Medications in GIO
Bisphosphonates

• **Alendronate Study**[109]
     o Population: Corticosteroid use >3months. Alendronate vs. placebo (12 month trial)
     o Spine BMD: +2-3% (Alendronate) vs. -0.4% (placebo)
     o Femoral neck BMD: +1% (Alendronate) vs. -1% (placebo)
     o 40% relative fracture risk reduction
     o Approved for treatment of GIO
 • **Risedronate Study**[110]
     o Population: Corticosteroid dose ≥ 7.5 mg/day x 6+ months (12 month trial)
     o Spine BMD: +2.7% (Risedronate) vs. no change in placebo

<ul><li>○ Femoral neck BMD: +1.8% (Risedronate) vs. no change in placebo</li><li>○ Relative risk of vertebral fracture decreased by 70%</li><li>○ Approved for treatment of GIO</li></ul><ul><li>**Zoledronic Acid vs. Risedronate Study**[111]<ul><li>○ Population: Patients starting or continuing prednisone ≥7.5mg/day (12 month trial)</li><li>○ Spine BMD: +4.1% (ZA) vs. +2.7% (Risedronate)</li><li>○ ZA had greater reduction of markers of bone turnover</li><li>○ ZA is approved for treatment of GIO</li></ul></li></ul>
Teriparatide
<ul><li>Consider as first line treatment in the following scenarios<ul><li>○ Severe osteoporosis (T-score <-3.5 or <-2.5 + fragility fracture)</li><li>○ Unable to tolerate/failure to respond to bisphosphonates</li></ul></li><li>**Teriparatide vs. Alendronate in patients with GIO**[112]<ul><li>○ Population: Corticosteroid dose ≥5mg for ≥3 months (18 month trial)</li><li>○ Spine BMD: +8.2% (Teriparatide) vs. +3.9% (Alendronate)</li><li>○ Total Hip BMD: +3.8% (Teriparatide) vs. +2.4% (Alendronate)</li><li>○ Femoral Neck BMD: +4.4% (Teriparatide) vs. +2.8% (Alendronate)</li></ul></li><li>**Teriparatide vs. Alendronate in patients with GIO**[113]<ul><li>○ Population: Corticosteroid dose ≥5mg for ≥3 months (36 month trial)</li><li>○ Spine BMD: +11.0% (Teriparatide) vs. +5.3% (Alendronate)</li><li>○ Femoral neck BMD: +6.3% (Teriparatide) vs. 3.4% (Alendronate)</li><li>○ New vertebral fractures: 3% (Teriparatide) vs. 13% (Alendronate)</li></ul></li></ul>
Denosumab
<ul><li>Not yet approved for use in glucocorticoid induced osteoporosis</li></ul>
Hormone (Testosterone, Estrogen)
<ul><li>Glucocorticoids reduce production of sex hormones</li><li>Consider as adjunct therapy</li></ul>

American College of Rheumatology Treatment Recommendations in GIO[114]

- Low risk patients (FRAX major osteoporosis fracture risk <10%)
    o No treatment if dose <7.5mg/day
    o Bisphosphonates if dose ≥7.5mg/day
- Medium risk patients (FRAX major osteoporosis fracture risk 10-20%)
    o Alendronate/Risedronate for any dose
    o Zoledronic acid if dose ≥7.5 mg/day
- High risk patients (FRAX major osteoporosis fracture risk >20%)
    o Bisphosphonates for any dose
    o Teriparatide for high risk patients taking ≥5 mg/day with duration <1 month or for any dose >1 month
- At risk premenopausal women and at risk men (<50 years old)
    o Bisphosphonates if dose ≥5mg for >1 month
    o Zoledronic acid if dose ≥7.5 mg/day
    o Zoledronic acid or Teriparatide if duration >3 months
- Premenopausal women with childbearing potential + high risk (previous fragility fracture)
    o Bisphosphonates if dose ≥7.5 mg/day for >3 months

Monitoring Response to Therapy

Prior to Initiation of Medication

- Obtain baseline DXA: 2 anatomic sites (hip, lumbar spine +/- distal third radius)
- Consider measuring bone turnover markers (NTX, CTX) if starting antiresorptive medication
    o Strongly consider in patients with malabsorption or questionable compliance
    o Obtain prior to initiation of treatment, and repeat at 3-6 months
    o Adequate response to treatment[76]
        - NTX: >50% decrease compared to baseline
        - CTX: >30% decrease compared to baseline

After Initiation of Medication

- Obtain DXA at 1 year following initiation of medication
- 2 anatomic sites (hip, lumbar spine +/- distal third radius)
- Preference is to obtain follow up DXA using same DXA scanner for accurate comparison to prior imaging

• BMD stable/improved→ repeat every 2 years • BMD decreased→ repeat yearly
After Discontinuation of Glucocorticoids
• Obtain DXA 1-2 years following discontinuation of glucocorticoids • If stable/improving BMD→ obtain follow up DXA every 5 years • Anticipate significant increase in BMD following cessation of exogenous glucocorticoids or reversal of endogenous Cushing's Syndrome
Drug Holiday
• Drug holidays are not recommended with continued glucocorticoid use

10 | Osteoporosis in Patients with CKD

Background Information
- It is estimated that greater than 10% of the US population is currently affected by chronic kidney disease[115]
- Providers must carefully differentiate osteoporosis from metabolic bone disease secondary to CKD, the latter of which becomes more prevalent in patients with eGFR <30 mL/min[116,117]
    - **Osteoporosis** is characterized by a decreased quantity of normally mineralized bone
    - **CKD-bone metabolic disease** is characterized by abnormal bone mineralization secondary to electrolyte and endocrine abnormalities and requires unique treatment considerations
- CKD-bone metabolic disease is associated with hypocalcemia, hyperphosphatemia, vitamin D deficiency, and secondary hyperparathyroidism[116]
- End-stage CKD with phosphorus retention, secondary hyperparathyroidism, and elevated acid loads is associated with increased risk of fragility fractures and fracture related mortality[116,118]
- Given the lack of sufficient evidence and consensus in the management of osteoporosis and CKD-bone metabolic disease in patients with eGFR <30 mL/min, **this section will focus on the management of patients with eGFR >30 mL/min and without evidence of CKD-bone metabolic disease**
- Patients with eGFR <30 mL/min or with evidence of CKD-bone metabolic disease are best managed by a bone metabolism expert

Clinical Manifestations
- No manifestations until fracture
- Consider clinical manifestations of CKD

Osteoporosis Screening in CKD

\	**Screening for Osteoporosis in CKD**
Imaging	- DXA (hip, lumbar spine +/- distal third radius)
- DXA has the same diagnostic value for patients with eGFR >30 mL/min[119,120]
- FRAX does not include any adjustment of risk for low eGFR
- Providers should consider adjusting the absolute fracture risk to a higher level in patients with CKD stages 3-5
Mandatory Labs	- CBC
- CMP: Creat/GFR, **bone specific alkaline phosphatase**, **phosphate**, **calcium** (with serum albumin for calculation of corrected serum calcium)
- **25-hydroxyvitamin D**
- **PTH**
- TSH
- 24-hour urine calcium
Lab Considerations	- +/- Tissue transglutaminase antibody (tTG) (symptoms of malabsorption/celiac)
- +/- Serum/urine protein electrophoresis (symptoms of multiple myeloma)
- +/- Salivary cortisol (symptoms of Cushing's syndrome)
- +/- Serum P1NP/CTX vs. urinary NTX to monitor response to therapy

Diagnostic Classification of Osteoporosis in CKD
- Must rule out CKD-bone metabolic disease
    o If eGFR >30 mL/min and no hypocalcemia, hyperphosphatemia, vitamin D deficiency, or secondary hyperparathyroidism→ classify osteoporosis as in postmenopausal osteoporosis
1) T-score <-2.5
2) Clinical presence of fragility fracture
3) T-score -1.0 to -2.5→ calculate FRAX score
    o Identification of high risk patients, as indicated by a FRAX 10 year hip fracture risk >3%, or a 10 year major osteoporotic fracture risk >20%

Pretreatment Evaluation
- Pretreatment evaluation should focus on differentiating osteoporosis from CKD-bone metabolic disease. If evidence of CKD-bone metabolic disease, refer to a bone metabolic specialist for appropriate management. If no evidence of CKD-bone metabolic disease is present, proceed with treatment as outlined.
- Traditional osteoporosis treatments (bisphosphonates, Denosumab, Teriparatide) have been shown to have similar efficacy in patients with CKD without evidence of CKD-bone metabolic disease[121,122]

Candidates for Pharmacologic Therapy
- Candidates for pharmacologic therapy should have no evidence of underlying CKD-bone metabolic disease as previously described.
1) T-score <-2.5
2) Clinical presence of fragility fracture
3) T-score -1.0 to -2.5→ FRAX 10 year hip fracture risk >3%, 10 year major osteoporotic fracture risk >20%

CKD Osteoporosis Treatment Protocol
Lifestyle Modifications
1) Weightbearing exercise program
2) Smoking cessation
3) Modest alcohol intake (<2 drinks daily)
4) Fall prevention
5) Calcium: **1200** mg/day (500mg or less provided by supplementation to reduce risk of arterial calcification and CVD) if eGFR >30 mL/min and no CKD-bone metabolic disease
6) Vitamin D: **800** IU daily if eGFR >30 mL/min and no CKD- bone metabolic disease
7) Osteoporosis specific medication (see chart below)

Osteoporosis Medication: CKD without CKD-Bone Metabolic Disease

First Line
- **Alendronate** (70mg po weekly) or **Risedronate** (35mg po weekly)
    - **Contraindicated if eGFR <30 mL/min**, though use in patients with more advanced CKD has been reported as well tolerated[123,124]
- **Zoledronic acid** (5mg IV yearly)→ if contraindication to oral

	bisphosphonates o **Contraindicated if eGFR <35 mL/min**
Second Line	
	• **Denosumab** (60mg SC q6mo)→ eGFR <30 mL/min, unresponsive to other therapies • **Teriparatide** (20mcg SC daily)→ severe osteoporosis + fracture, unresponsive to other therapies
Hormone	
	• **Premenopausal women with CKD + amenorrhea**: consider estrogen replacement for prevention, not first line for treatment[125] • **Men with CKD + symptomatic hypogonadism**: consider testosterone replacement

Monitoring Response to Therapy

Prior to Initiation of Medication
- Obtain baseline DXA: 2 anatomic sites (hip, lumbar spine +/- distal third radius)
- Monitoring markers of bone turnover (NTX, CTX) is less useful in patients with eGFR <30 mL/min, but can still be considered in patients with eGFR >30 mL/min without evidence of CKD-bone mineral disease[126]

After Initiation of Medication
- Obtain DXA at 2 year intervals following initiation of medication
- 2 anatomic sites (hip, lumbar spine, +/- distal third radius)
- Preference is to obtain follow up DXA using same DXA scanner for accurate comparison to prior imaging
- If eGFR <30 mL/min: monitor serum calcium, phosphorus, 25-hydroxyvitamin D, PTH levels every 4 months, along with serum creatinine every 12 months

Drug Holiday (Bisphosphonates)
- Given the long half-life of bisphosphonates, it is reasonable to consider a drug holiday for select patients
- Criteria: use of **Alendronate/Risedronate** x 5 years, or **Zoledronic Acid** x 3 years with a stable/improving BMD, no fragility fractures, and low risk for fracture in near future
- Obtain DXA 2 years following cessation of therapy
- Restart bisphosphonates with persistent bone loss (>5% on 2 subsequent DXAs 2 years apart) or if the patient sustains a fragility fracture during the drug holiday

11 Treatment: Calcium

Background Information
- Calcium is an essential nutrient that serves as a building block in the structural integrity of bone[127]
- The primary mineral component of bone is hydroxyapatite $Ca_{10}(PO_4)_6(OH)_2$
- Approximately 99% of the body's calcium stores reside in bone tissue[127]
- Serum calcium is maintained through dietary calcium intake
- When dietary intake is insufficient, the parathyroid glands release parathyroid hormone, which functions to increase serum calcium through a variety of mechanisms
- Parathyroid hormone increases serum calcium levels by stimulating bone resorption, increasing renal reabsorption of calcium and increasing excretion of phosphate, and increasing renal conversion of 25-hydroxyvitamin D to 1,25-dihydroxyvitamin D3, which promotes increased intestinal absorption of calcium[128]
- Prolonged periods of inadequate dietary intake of calcium can lead to decreased BMD and increased risk of fracture

Calcium Supplementation
Calcium Carbonate vs. Calcium Citrate[129]
- **Calcium carbonate**: absorption better with meals, poor with PPIs/H2 blockers - **Calcium citrate**: absorption equal with meals or fasting, better absorption with PPIs/H2 blockers compared to calcium carbonate
Calcium Dosing[130]
- Pre-menopausal dose: **1000 mg** daily (elemental) - Post-menopausal/male dose: **1200 mg** daily (elemental) - Maximum intake: Avoid >**2000 mg** daily - Calcium supplementation >**500 mg** daily should be given in divided doses - Half of calcium should ideally come from dietary sources   o Decreases risk of kidney stones, cardiovascular disease
Sources of Calcium
- 300mg/serving: 8oz. milk/yogurt, 1oz. hard cheese - 150mg/serving: 4 oz. cottage cheese/ice cream

	• 100-200mg/serving: dark green vegetables, nuts, breads, fortified cereals
Side Effects of Calcium Supplementation	

	• Kidney stones
	    • Supplemented calcium has increased risk compared to dietary calcium[131]
	• Cardiovascular disease
	    • Studies show increased risk of MI with supplemented calcium[132]
	    • Studies show a mild decreased risk of MI with dietary calcium[133]
	• Dyspepsia
	• Constipation
	• Malabsorption of iron and thyroid hormone

12 Treatment: Vitamin D

Background Information
- Vitamin D is primarily obtained through exposure to UVB sunlight radiation[134]
- UVB converts 7-dehydrocholesterol to previtamin D3 in the skin→ rapidly converts to vitamin D3→ metabolized to 25-hydroxyvitamin D in the liver by vitamin D-25-hydroxylase→ metabolized to active form of 1,25-dihydroxyvitamin D in the kidney by 25-hydroxyvitamin D-1α-hydroxylase[135]
- Vitamin D enhances intestinal absorption of calcium and phosphate. It may also improve proximal muscle strength and decrease fall risk in the elderly[136,137]
- Low serum vitamin D leads to impaired intestinal calcium and phosphate absorption[138]
- Low serum calcium triggers 1,25 dihydroxyvitamin D to travel to skeleton→ increased RANKL→ increased osteoclast function→ increased serum calcium at the expense of BMD[135,139]

- **Normal serum vitamin D** is defined as serum 25-hydroxyvitamin D between **30-50 ng/mL**[140]
- **Vitamin D deficiency** is defined as serum 25-hydroxyvitamin D **<20 ng/mL**[140]
- **Vitamin D insufficiency** is defined as 25-hydroxyvitamin D **21-29 ng/mL**[140]

- Over 41% of adults have 25-hydroxyvitamin D levels <20 ng/mL[141]
- Factors that influence vitamin D absorption include increased skin pigmentation, advancing age, geographic location, body habitus, and concurrent use of certain medications
- Other factors that contribute to vitamin D deficiency include low vitamin D intake, minimal sunlight exposure, and decreased synthesis (as seen in hepatic/renal disease)

Vitamin D Supplementation

Vitamin D3 (Cholecalciferol) vs. Vitamin D2 (Ergocalciferol)
- Vitamin D3 (cholecalciferol) and vitamin D2 (ergocalciferol) have a similar effect on serum vitamin D levels[142]

Vitamin D Dosing[143]
- High risk individuals with 25-hydroxyvitamin D <20 ng/mL
  - **50,000 IU** weekly x 6-8 weeks, then maintenance dose **1000 IU** daily
  - May need to continue 50,000 IU weekly for additional 6-8 weeks if patient remains vitamin D deficient on follow up testing
- High risk individuals with 25-hydroxyvitamin D 20-30 ng/mL
  - **800-1000 IU** daily
- Patients with malabsorption
  - **10,000-50,000 IU** daily may be required for patients with gastrectomy or malabsorption
- General Population: **800-1000 IU** daily
- Maximum intake: avoid >**4000 IU** daily unless medically necessary
- OK to take oral vitamin D fasting or with meals
- Consider Ultraviolet B sunlamp in vitamin D deficient patients refractory to oral supplementation

Sources of Vitamin D
- **Sunlight exposure (UVB)**
- Commercially fortified milk: 100 IU/8oz.
- Cod liver oil/fish oils
- Mushrooms exposed to sunlight

Side Effects of Vitamin D Supplementation
- Excessive vitamin D and calcium supplementation can lead to hypercalcemia, hypercalciuria, kidney stones[140]
- Chronically high vitamin D levels may increase risk of pancreatic cancer, prostate cancer, and increase mortality[144,145]

13   Treatment: Estrogen

Estrogen Pharmacology
- Estrogens are widely prescribed for the purposes of birth control, management of perimenopausal symptoms, and prevention/treatment of osteoporosis in post-menopausal women
- Common routes of administration include oral tablets and transdermal patches

Mechanism of Action
- Estrogen therapy has effects on multiple organ systems — below is a summary of those related to bone mineral density and osteoporosis
- Estrogen stimulates osteoblasts to release factors which decrease osteoclast activity and number[146]
- Estrogen increases osteoblast proliferation and function[146]
- Estrogen decreases maturation of osteoclasts by decreasing progenitor cell responsiveness to RANKL[146]
- Estrogen improves muscular strength and balance, which decreases risk of falls[147-149]
- Estrogen increases calcium absorption in the intestine[150]
- To summarize, estrogen therapy prevents bone loss and decreases risk of fragility fracture[151]

Indications for Estrogen
- Contraceptive
- Perimenopausal symptom management: vasomotor, urogenital
- **Prevention and treatment of osteoporosis** in select patients

Pretreatment Evaluation
- Must evaluate for contraindications to use, including a history of breast cancer, endometrial cancer or hyperplasia, DVT, CAD, and uncontrolled HTN
- Obtain baseline serum estrogen, LH/FSH
- Obtain baseline mammography

Adverse Effects of Estrogen
- **Breast cancer**: increased lifetime risk of breast cancer when estrogen is used in postmenopausal women ≥50 years old for a

duration of 5 years or longer. This increased risk appears to be related to lifetime duration of estrogen exposure, as use of estrogen in patients with early menopause is not associated with increased risk of breast cancer[152]
- **Endometrial cancer**: increased risk of endometrial cancer with prolonged use of unopposed estrogen therapy[153,154]
- **Thromboembolic disease**: increased risk of venous thromboembolism in patients taking oral estrogens[155]
- **Biliary disease**: increased incidence of cholelithiasis and cholecystitis in postmenopausal women taking estrogen[156]
- **Stroke**: increased risk of stroke in patients taking oral estrogens, which appears to be related to dose and duration of therapy[157]

Estrogen

Efficacy of Estrogen
- **Meta-analyses of Therapies for Postmenopausal Osteoporosis**[158]
    - Population: Postmenopausal women received estrogen +/- progestin vs. placebo (12+ months duration)
    - Spine BMD: +5.4% compared to placebo
    - Femoral neck BMD: +2.5% compared to placebo
    - Radius BMD: +3.0% compared to placebo
- **Women's Health Initiative Study**[159]
    - Population: Postmenopausal women 50-79yo, estrogen +/- progestin vs. placebo (5 year duration)
    - Estrogen alone
        - Spine fracture: 0.64 relative risk reduction
        - Hip fracture: 0.65 relative risk reduction
        - Total fracture: 0.71 relative risk reduction
    - Estrogen + Progestin
        - Spine fracture: 0.65 relative risk reduction
        - Hip fracture: 0.67 relative risk reduction
        - Total fracture: 0.76 relative risk reduction
- **Meta-analyses of HRT in Prevention of Vertebral and Nonvertebral Fractures**[160,161]
    - Population: Varied by study but generally included postmenopausal women with diagnosed osteoporosis, estrogen vs. placebo (12+ months duration)
    - Spine fracture: 33% decreased risk
    - All other fractures: 27% decreased risk

Dosing Estrogen

• **Million Women Study**[162]
     o Population: Women 50-64yo, various doses and routes of administration of estrogen
     o Reduction in fracture incidence associated with hormone therapy did not vary significantly between the various doses and routes of estrogen administration
     o Low dose estrogen is preferred as it appears to be effective in attenuating fracture risk and carries fewer side effects compared to higher doses of estrogen

Cessation of Estrogen Therapy
• Cessation of hormone therapy leads to a rapid decline of BMD and increased fracture risk within 1-2 years[162]
Estrogen Considerations
• No consensus currently exists regarding the use of hormone replacement therapy in the prevention and treatment of postmenopausal osteoporosis   • In postmenopausal women <60 years old with high fracture risk, it is reasonable to proceed with estrogen replacement in the prevention/treatment of osteoporosis after careful consideration of risks associated with estrogen use   • Add progestin for endometrial protection in women with a uterus   • Estrogens are contraindicated in patients with a history of breast cancer, endometrial cancer or hyperplasia, DVT, CAD, and uncontrolled HTN

14 Treatment: Bisphosphonates

Bisphosphonate Pharmacology
- Bisphosphonates are poorly absorbed: <1% via oral route (of which, 60% binds bone, 40% is renally cleared)[163-165]
- Bisphosphonates bind to the calcium-phosphorus surface on bone exposed during the resorption phase of bone remodeling[163,164]
- Bisphosphonates are not metabolized→ detached bisphosphonates either reattach at different bone sites, or are renally cleared[165]
- Alendronate, Risedronate, and Zoledronic Acid vary in affinity for bone surface and detachment rates[166]

Mechanism of Action
- Osteoclasts engulf bisphosphonates→ inhibition of intra-cellular farnesyl pyrophosphate synthesis→ decreases prenylation of osteoclast proteins necessary for bone resorption[166,167]
- Decreases bone resorption depth, osteoclast number, remodeling space[168]
- Decreases cortical porosity and perforation of trabecular plates[169,170]
- Increases trabecular bone volume and connectivity[169]

Indications for Bisphosphonates
- **Osteoporosis**
- Hypercalcemia
- Paget Disease
- Malignancies (multiple myeloma, breast cancer, prostate cancer)

Pretreatment Evaluation
- Hypocalcemia/vitamin D deficiency
    - Require correction prior to starting bisphosphonates
- Chronic kidney disease
    - Alendronate/Risedronate are contraindicated in patients with an eGFR <30 mL/min
    - Zoledronic acid is contraindicated in patients with an eGFR <35 mL/min
- Esophageal abnormalities
    - Avoid oral bisphosphonates in patients with Barrett's esophagus, esophageal dysmotility
- Bisphosphonate use immediately after fracture

- o Studies in humans have shown no difference in fracture healing time, rate of nonunion, or functional outcomes when starting bisphosphonates at various intervals immediately following fracture[171,172]
- o In patients already taking bisphosphonates, studies have shown patients have slightly longer duration until radiographic union with questionable clinical significance and no current recommendation exists to discontinue bisphosphonates following fracture given the long half-life of this drug class[173]
- o General recommendation is to initiate bisphosphonate therapy 4-6 weeks following acute fracture

Adverse Effects of Bisphosphonate Therapy
- **Atypical femur fractures (subtrochanteric fracture)[174]**
    - o Atypical femur fractures include spontaneous subtrochanteric and diaphyseal femur fractures that often present with prodromal thigh pain prior to fracture
    - o Rare: 3.2-50 in 100,000 patient years
    - o Most reported cases are in patients taking bisphosphonates for >5 years[175]
    - o Risk of atypical fracture decreases once bisphosphonates are stopped[175]
    - o Treatment includes protected weightbearing on the affected side, cessation of bisphosphonate +/- addition of PTH
    - o Prophylactic femoral intramedullary rodding if symptoms fail to improve in 2-3 months
- **Osteonecrosis of the jaw (ONJ)[176]**
    - o ONJ is exposed bone in the maxillofacial region that does not heal within 8 weeks
    - o Rare: <1 in 100,000 patient years
    - o Increased risk with IV bisphosphonates, cancer, greater duration of exposure, dental extractions/implants, glucocorticoids, smoking, pre-existing dental disease
- GI (reflux, esophagitis, ulcers)
    - o Risk is very low if taken properly → OK to give in patients with well controlled GERD
- Esophageal/gastric cancer
    - o Some studies show a small increased risk in patients who have taken oral bisphosphonates for ≥5 years duration[177]

- Flu-like symptoms[178]
    - Common with first infusion of ZA: 32%
    - Decreased risk with subsequent infusions (6.6% second infusion, 2.8% third infusion)
    - Symptoms generally begin within 24-36 hours and last 3 days
    - Patients can reduce symptoms by taking OTC Acetaminophen during this timeframe
- Hypocalcemia[179]
    - IV bisphosphonates pose greater risk than oral
    - More profound in patients with vitamin D deficiency
    - If 25-hydroxyvitamin D is <20 ng/mL→ supplement vitamin D until levels >25-30 ng/mL before initiating bisphosphonate therapy
- Acute renal failure (ARF)[180]
    - There are reports of ARF in patients on IV bisphosphonates with multiple myeloma, diuretic use, NSAIDs
- Atrial fibrillation[181]
    - Atrial fibrillation has been reported after initiation of bisphosphonate therapy, however, a direct causal effect has not been substantiated
    - If atrial fibrillation develops, it is generally transient and there is no increased risk of MI or stroke
- Ocular side effects
- Musculoskeletal pain

Alendronate (Fosamax)

Dosing Alendronate
- Daily (10mg) oral vs. weekly (70mg) oral

Efficacy of Alendronate
- **Fracture Intervention Trial (FIT)**[182]
  - Population: Postmenopausal women, + vertebral fracture, FN BMD <-2.1. Alendronate vs. placebo (3 years duration)
  - Femoral neck BMD increase: 4.1% → 30% decreased risk of fracture
  - Spine BMD increase: 6.2% → 50% decreased risk of fracture
- **Fracture intervention Trial (FIT)**[183]
  - Population: Postmenopausal women, - vertebral fracture, FN BMD <-2.5. Alendronate vs. placebo (4 years duration)
  - Femoral neck: 56% decreased risk of fracture
  - Spine: 44% decreased risk of fracture
  - All clinical fractures: 36% decreased risk of fracture
- Long-term efficacy[184]
  - Well tolerated for at least 10 years
  - 10-year increase in spine BMD: 13.7% (10% in first 5 years)

Drug Holiday Considerations
- Consider discontinuation of Alendronate after 5 years if stable BMD, no fragility fractures
- **Fracture Intervention Trial Long-Term Extension (FLEX)**[184]
  - Population: **FIT** trial patients switched to placebo after 5 years of Alendronate vs. continuation of Alendronate (5 years duration)
  - Hip BMD decline in placebo group: 2.4%
  - Spine BMD decline in placebo group: 3.7%
  - Increased markers of bone turnover in placebo group
  - Patients with hip T-score <-2.5: increased risk of fracture in placebo group compared to continuation of Alendronate group
  - Increased risk of clinical vertebral fractures
  - Mean BMD in placebo group remained higher than levels 10 years prior
  - No ONJ in those treated with Alendronate for 10 years
  - Highest risk of fracture (T-score <-3.5) were excluded

from trial→ recommend continuation of Alendronate for 10 years total

Risedronate (Actonel)

Dosing Risedronate
- Daily (5mg) oral vs. Weekly (35mg) oral vs. Monthly (150mg) oral
- Enteric-coated tablets available

Efficacy of Risedronate
- **Vertebral Efficacy with Risedronate (VERT)**[185]
  - Population: Postmenopausal women, 2+ vertebral fractures or 1 vertebral fracture + spine BMD <-2.0, Risedronate vs. placebo (3 years duration)
  - Spine BMD: 5.4% increase → 41% decreased fracture risk
  - Femoral neck BMD: 1.6% increase → 39% decreased fracture risk
  - Trochanter BMD: 3.3% increase→ 39% decreased fracture risk
- Reduced risk of hip fracture in patients with confirmed osteoporosis, but not in patients selected on basis of risk factors alone
- Long-term efficacy[186]
  - Well tolerated for at least 7 years

Drug Holiday Considerations
- **VERT Extension Study**[187]
  - Population: **VERT** trial patients (3 years Risedronate) followed by 1 year drug holiday
  - Spine BMD: 0.83% decrease
  - Hip BMD: 1.23% decrease
  - Markers of bone turnover returned to baseline at 1 year
  - New vertebral fracture risk remained 46% reduced
  - Recommendation is treatment of Risedronate and Alendronate the same when considering drug holidays

Zoledronic Acid (Reclast)

Dosing Zoledronic Acid
- Once yearly 5mg IV dosing

Efficacy of Zoledronic Acid
- **Health Outcomes and Reduced Incidence with Zoledronic Acid Once Yearly (HORIZON)**[188]
  - Population: Postmenopausal women, FN T-score <−2.5 or T-score <−1.5 + XR evidence of 2 mild vertebral compression fractures, or 1 moderate vertebral compression fracture. Zoledronic Acid vs. placebo (3 years duration)
  - Spine: 70% decreased fracture risk
  - Femoral neck: 41% decreased fracture risk
- Long-term efficacy
  - Well tolerated for 3-6 years

Drug Holiday Considerations
- **HORIZON extension study**[189]
  - Population: **HORIZON** trial patients, discontinuation of ZA after 3 years vs. continuation for 3 additional years (3 years duration)
  - Spine BMD: +0.24% (ZA)   −0.80% (placebo)
  - Hip BMD: +3.2% (ZA)   +1.18% (placebo)
  - No difference in incidence of nonvertebral, clinic vertebral, or all clinical fractures
  - Nonsignificant increases in side effects observed in patients on ZA for 6 years
  - Beneficial effects of ZA continue for 3 years following 3 years of therapy

Monitoring Response to Therapy
- Adequate treatment response is defined by stable/improving BMD at 2 year follow up
- Stable or improved BMD→ obtain follow up DXA in 2 years
- Decline in BMD <5% → obtain follow up DXA in 1 year
- Decline in BMD >5% → change from oral to IV bisphosphonate, Teriparatide, or Denosumab
- Switching to Teriparatide or Denosumab should be considered for patients with severe osteoporosis (T-score <-2.5 + fragility fracture who continue to fracture on bisphosphonates)

15 Treatment: Denosumab (Prolia)

Denosumab Pharmacology
- To understand how Denosumab increases BMD and decreases fracture risk, one must understand the RANK-RANKL-OPG Pathway and how it relates to bone resorption
- RANKL is secreted by osteoblasts and interacts with RANK on osteoclast precursor cells
- RANKL/RANK interaction results in differentiation of osteoclasts and activates bone resorption
- Osteoprotegerin (OPG) is also secreted by osteoblasts→ OPG binds RANKL and inhibits its function, resulting in decreased bone resorption[190]

Mechanism of Action
- Denosumab is a human monoclonal antibody to receptor activator of nuclear factor Kappa-B ligand (RANKL)[191]
- Denosumab binds RANKL→ blocks binding of RANKL to RANK
- Denosumab inhibits osteoclast formation, decreases bone resorption, increases BMD
- **Improvement in cortical thickness, bone mineral content, porosity, and strength**[192]
- Denosumab is not renally excreted and no dosing adjustments are required for patients with CKD

Indications for Denosumab
- Men and postmenopausal women with osteoporosis
- Men with osteoporosis secondary to ADT[193]
- Hypercalcemia of malignancy[194]

Pretreatment Evaluation
- Must correct hypocalcemia and vitamin D deficiency prior to starting Denosumab
- Caution with use in patients taking immunosuppressants[195,196]

Adverse Effects of Denosumab Therapy
- **Most common**: musculoskeletal pain, hypercholesterolemia, cystitis, eczema, cellulitis

- **Hypocalcemia**
    - Transient asymptomatic hypocalcemia has been observed in patients given Denosumab[197]
    - More common in patients with CKD, malabsorption, hypoparathyroidism
    - Do not give in patients with pre-existing hypocalcemia until calcium level is corrected
- **Osteonecrosis of the Jaw/Atypical Femur Fractures**
    - Very low risk→ similar to that observed in patients with bisphosphonate use
- **Immune System**
    - RANKL also has functions within the immune system as it is expressed by lymphocytes and other cells involved in the immune response[195]
    - Some studies show an increased number of patients with infections requiring hospitalization, however, no statistically significant decrease in WBCs has been identified in patients on Denosumab[196]

Denosumab
Dosing Denosumab
• 60mg subcutaneous injection every 6 months
Efficacy of Denosumab: Postmenopausal Women

• **FREEDOM Trial**[197]
     ○ Population: Postmenopausal women 60-90yo, T-score -2.5 to -4.0 (3 years duration)
     ○ Spine BMD: +9.2% (Denosumab) vs. 0% (placebo)
         ▪ Spine fractures: 2.3% vs. 7.2% respectively
     ○ Total Hip BMD: +6.0% (Denosumab) vs. 0% (placebo)
         ▪ Hip fractures: 0.7% vs. 1.2% respectively
     ○ Decreased markers of bone turnover with Denosumab
 • **FREEDOM Extension Trial**[198]
     ○ Population: **FREEDOM** trial patients received additional 3 years of Denosumab
     ○ Spine BMD: +15.2% total
     ○ Total hip BMD: +7.5% total
 • **Phase II trial: 60mg subcutaneous x 4 years**[199]
     ○ Population: Postmenopausal women, T-score -1.8 to -4.0 received Denosumab 60mg SC q 6 months vs. Alendronate vs. placebo (4 years duration)
     ○ Denosumab spine BMD: +9.4-11.8%

- o   Denosumab total hip BMD: +4.0-6.1%
- **Phase II trial: 60mg subcutaneous x 8 years**[200]
  - o   Denosumab spine BMD: +16.5%
  - o   Denosumab total hip BMD: +6.8%
  - o   Denosumab distal radius BMD: +1.3%
  - o   Adverse effects: no atypical femur fractures, ONJ with 8 years of treatment
- **Phase III Pivotal Fracture Study**[197]
  - o   Population: Denosumab 60mg q 6 months in postmenopausal women (60-91yo) with osteoporosis vs. placebo (3 years duration)
  - o   Spine fracture: 68% reduced risk
  - o   Hip fracture: 40% reduced risk
  - o   Nonvertebral fracture: 20% reduced risk

Efficacy of Denosumab: Men

- **Denosumab in men with low BMD**[201]
  - o   Population: Men (mean age 65) with low BMD received Denosumab 60mg q 6 months vs. placebo (12 month duration)
  - o   Femoral neck BMD: +2.1% (Denosumab) vs. +0% (placebo)
  - o   Total hip BMD: +2.4% (Denosumab) vs. +0.3% (placebo)
  - o   Spine BMD: +5.7% (Denosumab) vs. +0.9% (placebo)

Monitoring Considerations

- Monitor for the development of hypocalcemia
  - o   At risk populations include patients with an eGFR <30 mL/min, hypoparathyroidism, malabsorption syndromes
  - o   **Monitor serum calcium 10 days after starting therapy**
- Monitor closely for signs of infection, particularly in immunocompromised patients
- Follow up DXA 2 years following initiation of therapy

Discontinuation Considerations

- Effects of Denosumab on BMD are reversible with discontinuation
- Cessation of Denosumab therapy requires switching to an alternative osteoporosis medication
- **Phase II trial: 60mg subcutaneous x 2 years**[199]
  - o   Population: Postmenopausal women (T-score -1.8 to -4.0) received Denosumab 60mg SC q 6 months x 24 months, then no treatment x 24 months
  - o   Spine BMD following cessation: -6.6%
  - o   Total hip BMD following cessation: -5.3%

- Bone turnover markers: increased within 3-6 months of discontinuation

16 Treatment: Teriparatide (Forteo)

Pharmacology of Teriparatide
- Teriparatide is a recombinant form of human parathyroid hormone
- Teriparatide is identical to the biologically active 34 N-terminal amino acid sequence of the 84-amino acid parathyroid hormone[202]
- The primary function of parathyroid hormone is maintenance of ionized serum calcium concentrations, though it also effects the regulation of serum phosphorus and vitamin D synthesis
- Teriparatide is self-administered daily via subcutaneous injection
- Teriparatide is approved for 24 months of continuous therapy

Mechanism of Action
- Unlike the antiresorptive class of osteoporosis medications which decrease bone turnover, Teriparatide increases bone turnover and preferentially increases the formation of new bone
- PTH stimulates an increase in the number of osteoblasts in cancellous and periosteal bone[203]
- Osteoblasts release cytokines that activate osteoclasts, increasing bone turnover
- Early in the course of treatment, bone formation outpaces bone resorption[204,205]
- New bone formation peaks at 6 months following initiation of Teriparatide[204,205]
- With prolonged exposure to PTH, there is a reestablishment of equilibrium between bone formation and resorption leading to a plateau in new bone acquisition which occurs between 18-24 months[204-206]
- Overall, Teriparatide positively impacts bone size, microarchitecture, and rejuvenates bone matrix[207]

Chronic Hyperparathyroidism vs. Intermittent Exogenous PTH Administration
- Intermittent PTH increases OPG and decreases RANKL[208]
- Intermittent PTH causes increased osteoblast activity→ deposition of bone without previous resorption[209]
- Intermittent PTH inhibits sclerostin expression→ less inhibition of bone formation[210]
- Intermittent PTH increases IGF-1 and other anabolic growth factors at tissue level[211]

Indications for Teriparatide
- **Osteoporosis**
- Off label use of Teriparatide to aid in fracture healing, osteonecrosis of the jaw, and endocrine derangements has been studied with successful outcomes[212]

Candidates for Therapy
- Teriparatide as a first line treatment
    - Patients with severe osteoporosis
        - Men/post-menopausal women with T-score <-3.5
        - Men/post-menopausal women with multiple prior fragility fractures
    - Osteoporosis patients who cannot tolerate or have contraindications to bisphosphonates
- Teriparatide as a second line treatment
    - Patients who fail other osteoporosis therapies (ongoing BMD loss, new fragility fracture)
- Avoid in patients at high risk for osteosarcoma, including patients with a history of Paget disease, radiation treatments, and unexplained elevated bone specific alkaline phosphatase

Pretreatment Evaluation
- Must correct hypocalcemia and vitamin D deficiency prior to starting Teriparatide
- If uric acid levels are >7.5 mg/dL, one must weigh risks and benefits of Teriparatide
- Avoid in patients at high risk for osteosarcoma

Adverse Effects of Teriparatide Therapy
- **Most common**: nausea, headache, muscle cramps
- **Hypercalcemia**
    - Transient, asymptomatic hypercalcemia occurs in approximately 5% of patients[213]
    - Case studies have reported more persistent hypercalcemia requiring hospitalization in select patient populations[214]
- **Hypercalciuria**
    - Mild increased urinary calcium has been observed in patients on Teriparatide therapy[215]
    - This is of questionable clinical significance and routine monitoring of urinary calcium is not recommended

- **Hyperuricemia**
    - Teriparatide may elevate uric acid levels in select patient populations, which can precipitate gout[212]
- **Osteosarcoma**
    - Osteosarcoma has been identified in rat models exposed to very high doses of Teriparatide for durations similar to the lifespan of the rat[216]
    - Teriparatide does not appear to increase risk of osteosarcoma in human populations

Teriparatide (Forteo)
Dosing Teriparatide
• 20mcg subcutaneous daily x 2 years
Efficacy of Teriparatide

• **Fracture Prevention Trial (FPT)**[213]
  ○ Population: Postmenopausal women, + vertebral fracture, 20 vs. 40 mcg/day dosing vs. placebo (18 month duration)
  ○ 20 mcg/day vs. placebo
   ▪ Spine BMD: +9% vs. placebo
   ▪ Femoral neck BMD: +3% vs. placebo
   ▪ New vertebral fracture: 5% vs. 14% placebo
  ○ 40 mcg/day vs. placebo
   ▪ Spine BMD: +13% vs. placebo
   ▪ Femoral neck BMD: +6% vs. placebo
   ▪ New vertebral fracture: 4% vs. 14% placebo
• **PTH in Men**[217]
  ○ Population: Men with T-score <-2.0, 20 vs. 40 mcg/day Teriparatide (11 month duration)
  ○ 20 mcg/day vs. placebo
   ▪ Spine BMD: +5.9% vs. placebo
   ▪ Femoral neck BMD: +1.5% vs. placebo
  ○ 40 mcg/day vs. placebo
   ▪ Spine BMD: +9% vs. placebo
   ▪ Femoral neck BMD: +2.9% vs. placebo
• **PTH vs. Alendronate in GIO**[218]
  ○ Population: Men and women with GIO, glucocorticoid dose ≥5mg/day for 3+ months, Teriparatide 20 mcg/daily vs. Alendronate 10mg daily (18 month duration)
  ○ Spine BMD: +7.2% (Teriparatide) vs. +3.4% (Alendronate)
  ○ Vertebral fracture: 0.6% (Teriparatide) vs. 6.1%

(Alendronate)

Monitoring Considerations

- **Serum calcium**: Consider monitoring (1, 6, 12 months)
    - Measure calcium 24 hours after last dose
    - If elevated—discontinue vitamin D and Calcium supplementation
- **Uric acid**: consider monitoring at baseline and 6 months
- **Bone markers**
    - Consider monitoring P1NP at start of therapy and at 6 months
    - P1NP + ≥21% or increase >10 mcg/L from baseline indicates adequate response[219]
- **DXA**: every 2 years

Discontinuation Considerations

- **Parathyroid Hormone and Alendronate for Osteoporosis (PaTH)**[220]
    - Population: Postmenopausal women received 12 months of PTH followed by 12 months of Alendronate vs. placebo
    - Lumbar spine BMD: +12.1% (PTH + Alendronate) vs. +4.1 (PTH + placebo)
- Teriparatide is approved for 24 months of continuous use
- Following this period, BMD and fracture risk will return to baseline over a 2 year period if the patient is not transitioned to an alternative osteoporosis medication[221]
- Consider transition to antiresorptive medications (Bisphosphonates, Denosumab)

17 Fragility Fracture Overview and Management

Hip Fractures: Femoral Neck and Intertrochanteric Fractures

Background Information
- **Femoral neck fractures** are located between the femoral head and the trochanters of the proximal femur
    - Intracapsular fractures with a tenuous blood supply
    - High risk of nonunion and avascular necrosis with displaced fractures
- **Intertrochanteric fractures** are located between the greater and lesser trochanter of the proximal femur
    - Extracapsular fractures with a rich blood supply
    - Low risk of nonunion, avascular necrosis
- **Mechanism of injury:** fall onto the lateral aspect of the hip

Classification[222]
- **Femoral neck fractures**
    - Stable fractures are entirely nondisplaced, or have a valgus impacted position
    - Unstable fractures are displaced to any degree
- **Intertrochanteric fractures**
    - Stable fractures have an intact posteromedial cortex
    - Unstable fractures have comminution/displacement of the posteromedial cortex, subtrochanteric extension, or reverse obliquity

Examination
- Inability to bear weight on affected hip
- Groin pain
- Leg positioning: leg shortened, externally rotated
- Careful neurovascular examination is warranted

Imaging
- XR: AP pelvis, AP + cross table lateral hip
- Consider CT/MRI if fracture is suspected but initial x-rays are equivocal

Treatment[222]
- The general recommendation for operative hip fractures is surgery

- within 48 hours
- Select patients, including nonambulators and those that are poor surgical candidates can be managed nonoperatively with extended nonweightbearing and close observation
- **Femoral neck fractures**
  - Treatment is surgical with closed reduction percutaneous pinning vs. intramedullary nailing for stable fractures, and hip hemiarthroplasty vs. total hip arthroplasty for unstable fractures
- **Intertrochanteric fractures**
  - Treatment is surgical with a sliding hip compression screw and lateral side plate vs. intramedullary nailing
  - Early weightbearing and physical rehabilitation is preferred

Distal Radius Fractures

Background Information

- Distal radius fractures involve a fracture to the distal third of the radius and can be associated with ulnar styloid fractures
- There is a rich vascular supply with a very low risk of nonunion and avascular necrosis
- The articular surface of the distal radius has an 11 degree volar tilt when viewed on the lateral XR, which is an important consideration when determining fracture displacement
- **Mechanism of injury:** fall onto outstretched hand (FOOSH)

Classification[222]

- **Colles (>90% of all distal radius fractures)**
  - Distal radius fracture with dorsal angulation
- **Smith (reverse Colles)**
  - Distal radius fracture with volar angulation
- **Barton**
  - Fracture dislocation of wrist→ dorsal or volar rim of distal radius is displaced
- **Chauffeur's (radial styloid fracture)**
  - Distal radius fracture involving the radial styloid
  - Can be associated with scapholunate ligament injuries

Examination

- Tenderness (+/- crepitus) with palpation over the distal radius, ulnar styloid

- Visible deformity of the wrist may be present → most commonly the wrist is dorsally angulated and radially deviated
- Evaluate neurovascular structures for compromise → attention to the median nerve for evaluation of acute carpal tunnel syndrome
- Evaluate flexor and extensor tendon function

Imaging
- Wrist XR: PA, lateral, oblique
- Consider CT scan for complex fracture patterns

Acceptable Fracture Alignment Parameters[222]
- Intra-articular incongruity: <1-2mm
- Radial shortening: <5mm
- Radial inclination: ≥15 degrees
- Dorsal tilt: <15 degrees
- Volar tilt: <20 degrees

Treatment[222]
- Stable/acceptable fracture alignment
    - Immobilization: short arm cast x 4 weeks, removable brace for additional 2-4 weeks
    - Range of motion: finger motion immediately to prevent stiffness, then progressive wrist ROM as tolerated following XR evidence of callus and cast removal
    - Weightbearing: minimal weightbearing x 6 weeks, then progressive WBAT once XR evidence of callus and clinical healing
- Unstable/unacceptable fracture alignment
    - Operative management is indicated
    - Treatment options: closed reduction + casting, closed reduction + percutaneous pinning, open reduction internal fixation with volar plate, or an external fixator

Proximal Humerus Fractures

Background Information
- Proximal humerus fractures involve a fracture to the proximal aspect of the humerus from the humeral head to the pectoralis major insertion
- Anatomically, the proximal humerus is divided into the head, shaft, greater tuberosity, and lesser tuberosity
- **Mechanism of injury:** fall onto outstretch hand vs. direct trauma

Classification[222]
<ul><li>The Neer classification system is most commonly used to classify proximal humerus fractures</li><li>The proximal humerus is divided into 4 anatomic regions (head, shaft, greater tuberosity, lesser tuberosity)</li><li>A part is defined as a fracture that is displaced >1cm or >45 degrees of angulation</li><li>Regardless of the number of fracture lines, a proximal humerus fracture is considered one part (nondisplaced) if no parts exceed 1cm displacement or 45 degrees angulation</li></ul>
Examination
<ul><li>Patients will brace the injured arm across their body with the contralateral hand</li><li>Careful neurovascular examination, with particular attention to axillary nerve</li><li>Deltoid atony from axillary nerve palsy can develop— typically resolves within 1-2 months</li></ul>
Imaging
<ul><li>XR: AP, axillary, lateral Y views</li><li>Consider CT scan for complex fractures</li></ul>
Acceptable Fracture Alignment Parameters
<ul><li>Stable fractures are nondisplaced as defined by the Neer classification system</li><li>Unstable fractures are displaced as defined by the Neer classification system</li></ul>
Treatment[222]
<ul><li>Stable/acceptable fracture alignment<ul><li>Immobilization: sling vs. immobilizer for 4-6 weeks</li><li>Range of motion: hand/wrist/elbow motion immediately. Passive shoulder ROM begins at 2-4 weeks, active shoulder ROM at 4-6 weeks</li><li>Weightbearing: Minimal weightbearing x 6 weeks, followed by progressive weightbearing as tolerated once XR evidence of callus and clinical healing</li></ul></li><li>Unstable/unacceptable fracture alignment<ul><li>Operative management is indicated</li><li>Treatment: closed reduction percutaneous pinning, open reduction internal fixation, intramedullary rodding, hemiarthroplasty, or total shoulder arthroplasty</li></ul></li></ul>

Pelvic Fractures: Pubic Rami and Sacral Insufficiency Fractures

Background Information
- **Pubic rami fractures** involve fractures to the superior and inferior pubic rami, which can be associated with other fractures/dislocations of the pelvis ring
- **Sacral fractures** involve fractures to the sacrum, which is formed by the fusion of the 5 sacral vertebrae and lies between the lumbar spine and the coccyx
- **Mechanism of injury:** fall onto hip or buttocks

Classification[222]
- Pubic rami fractures are classified by the orientation of the fracture, degree of displacement, and mention of associated pelvic ring fractures if present
- Sacral fractures are classified according to the Denis classification system
  - Zone 1: Fracture lateral to foramina
  - Zone 2: Fracture through foramina
  - Zone 3: Fracture medial to foramina

Examination
- Patients present with pelvic and groin pain
- Inability to bear weight or painful weightbearing
- Careful neurovascular examination is critical in evaluating fractures to the pelvic ring
- Evaluate for lower extremity neurovascular compromise, urinary/rectal/sexual deficits
- Evaluate pelvic ring for stability

Imaging
- XR: AP pelvis, inlet/outlet, cross table lateral
- CT scan if XRs are equivocal and high suspicion of pelvic fracture

Treatment[222]
- **Sacral insufficiency fractures**
  - Less than 1cm displacement and no neurologic compromise can be managed nonoperatively with analgesia and protected weightbearing
  - Greater than 1cm of displacement or neurovascular compromise requires operative intervention
- **Pubic rami fractures**
  - Stable pubic rami fractures without associated pelvic ring or neurovascular injuries can be managed nonoperatively

> with analgesia and protected weightbearing
> - Unstable pubic rami fractures and those associated with an unstable pelvic ring or neurovascular compromise may require operative intervention

Vertebral Compression Fractures

Background Information
- **Vertebral compression fractures** involve acute fractures of the lumbar and thoracic vertebrae
- Only one-third of vertebral compression fractures are symptomatic
- Vertebral compression fractures can lead to kyphotic posture and compromise pulmonary function
- **Mechanism of injury:** fracture may occur secondary to a low trauma fall, or spontaneously

Classification[223]
- Grade 0: <20% loss of height
- Grade 1: 20-25% loss of height, 10-20% reduction of vertebral area
- Grade 2: 25-40% loss of height, 20-40% reduction in vertebral area
- Grade 3: >40% loss of height, >40% reduction in vertebral area

Examination
- Focal tenderness with palpation of the spinous process of involved vertebrae
- Evaluate for neurovascular compromise

Imaging
- XR: AP and Lateral spine
- CT and MRI can be considered if XRs are nondiagnostic or if evidence of neurologic impairment

Treatment[222]
- The majority of uncomplicated vertebral compression fractures can be managed nonoperatively with analgesia, activity modification, +/- TLSO brace
- Vertebroplasty and kyphoplasty can be considered in patients with persistent severe pain despite compliance with nonoperative treatment. Must confirm the source of pain is from an acute compression fracture as other causes of acute back pain can co-exist in the setting of an acute compression fracture.
- Surgical decompression/stabilization may be indicated with evidence of acute neurologic compromise

18   Patient Osteoporosis Questionnaire

- At what age did you go through menopause (if applicable)?
- Personal history of low trauma fracture (sustained from a fall from standing height)?
- Personal history of stress fracture?
- Family history of osteoporosis or low trauma fracture?
- Number of falls in last year?
- History of smoking or use of tobacco products?
- Do you consume more than 2 alcoholic beverages daily?
- Have you ever been diagnosed with high or low calcium/vitamin D?
- Do you currently supplement calcium or vitamin D? If so, at what doses?
- Do you currently weigh <127 pounds?
- Have you ever been diagnosed with any of the following medical conditions?
    - Hyperthyroidism
    - Hyperparathyroidism
    - Celiac disease
    - Hypercalciuria (elevated calcium in the urine)
    - Multiple myeloma
    - Cushing's disease
    - Hypogonadism
    - Kidney or liver disease
    - Rheumatoid arthritis
    - Diabetes: Type I or II
- Have you taken any of the following medications?
    - Aromatase Inhibitors: Anastrazole, Letrozole, Exemestrane
    - Antidepressants: Zoloft, Celexa, Prozac
    - Anti-epileptic Drugs: Carbamazepine, Valproate, Phenobarbital, Phenytoin
    - Calcineurin Inhibitors: Cyclosporine, Tacrolimus
    - Depot medroxyprogesterone acetate
    - Glucocorticoids: Prednisone
    - GnRH Agonists: Leuprorelin, Triptorelin, Goserelin,
    - Proton Pump Inhibitors: Protonix, Nexium, Prevacid
    - Thiazolidinediones: Avandia

19 | Treatment of Osteoporosis: Patient Handout

1) **Smoking Cessation**

2) **Avoid excessive alcohol consumption of greater than 2 drinks per day**

3) **Weightbearing Exercise**
   - Weightbearing exercises have been shown to improve bone strength, improve muscle strength, and decrease the risk of falls
   - You should participate in weightbearing exercises for a minimum of 30 minutes three times weekly
   - Exercises include walking, hiking, dancing, yoga, golf, racquet sports, and strength training

4) **Calcium**
   - Calcium is an essential mineral that is needed to build strong bones
   - Dosing Requirements
     - Pre-menopausal women: **800-1000** mg daily
     - Post-menopausal women and men: **1200** mg daily
     - Maximum intake: avoid consuming more than 2000 mg daily
     - Half of daily calcium intake should come from dietary sources
   - Calcium supplements come in two forms: calcium carbonate and calcium citrate
     - Calcium carbonate should be taken with food, while calcium citrate can be taken with food or fasting
     - Calcium citrate is preferred if you take proton pump inhibitors or other acid reducers
   - Dietary sources of calcium
     - 300 mg per serving: 8 oz. milk/yogurt, 1 oz. hard cheese
     - 150 mg per serving: 4 oz. cottage cheese/ice cream
     - 100-200 mg per serving: dark green vegetables, nuts, breads, fortified cereals

5) **Vitamin D**
   - Vitamin D enhances your body's ability to absorb calcium and build strong bones

- Dosing Requirements
    - Pre-menopausal women: **600** IU daily
    - Post-menopausal women and men: **800-1000** IU daily
    - Maximum intake: avoid consuming more than 4000 IU daily
- Vitamin D comes in two forms: Vitamin D3 (cholecalciferol) and Vitamin D2 (ergocalciferol)
    - Both forms of Vitamin D can be taken fasting or with meals
    - Both Vitamin D3 and Vitamin D2 are well absorbed by the body and neither is preferred
- Sources of Vitamin D
    - Commercially fortified milk: 100 IU/8oz.
    - Cod liver oil/fish oils
    - Mushrooms exposed to sunlight
    - Sunlight exposure (UVB)

References

1. Cooper, C., Campion, G., & Melton III, L. (1992). Hip fractures in the elderly: a world-wide projection. *Osteoporosis international*, *2*(6), 285-289.
2. Baxter-Jones, A. D., Faulkner, R. A., Forwood, M. R., Mirwald, R. L., & Bailey, D. A. (2011). Bone mineral accrual from 8 to 30 years of age: an estimation of peak bone mass. *Journal of bone and mineral research*, *26*(8), 1729-1739.
3. General, S. (2004). Bone health and osteoporosis: a report of the surgeon general. *US Department of Health and Human Services, Office of the Surgeon General, Rockville, MD*.
4. Melton, L. J., Chrischilles, E. A., Cooper, C., Lane, A. W., & Riggs, B. L. (2005). How many women have osteoporosis?. *Journal of bone and mineral research*, *20*(5), 886-892.
5. Randell, A., Nguyen, T. V., Lapsley, H., Jones, G., Kelly, P. J., & Eisman, J. A. (1995). Direct clinical and welfare costs of osteoporotic fractures in elderly men and women. *Osteoporosis International*, *5*(6), 427-432.
6. Johnell, O., & Kanis, J. A. (2006). An estimate of the worldwide prevalence and disability associated with osteoporotic fractures. *Osteoporosis international*, *17*(12), 1726-1733.
7. Gullberg, B., Johnell, O., & Kanis, J. A. (1997). World-wide projections for hip fracture. *Osteoporosis international*, *7*(5), 407-413.
8. Marshall, D., Johnell, O., & Wedel, H. (1996). Meta-analysis of how well measures of bone mineral density predict occurrence of osteoporotic fractures. *Bmj*, *312*(7041), 1254-1259.
9. Kanis, J. A., Johnell, O., De Laet, C., Johansson, H., Odén, A., Delmas, P., & McCloskey, E. V. (2004). A meta-analysis of previous fracture and subsequent fracture risk. *Bone*, *35*(2), 375-382.
10. Riggs, B. L., & Melton III, L. J. (1986). Involutional osteoporosis. *New England journal of medicine*, *314*(26), 1676-1686.
11. Khosla, S., Riggs, B. L., Atkinson, E. J., Oberg, A. L., McDaniel, L. J., Holets, M., & Melton, L. J. (2006). Effects of sex and age on bone microstructure at the ultradistal radius: a population-based noninvasive in vivo assessment. *Journal of Bone and Mineral Research*, *21*(1), 124-131.
12. Gallagher JC: Osteoporosis. Conn's Current Therapy. Robert ER (ed). Philadelphia, WB Saunders Co, 1999, pp 590-594
13. Ryan, C. S., Petkov, V. I., & Adler, R. A. (2011). Osteoporosis in men: the value of laboratory testing. *Osteoporosis International*, *22*(6), 1845-1853.
14. Painter, S. E., Kleerekoper, M., & Camacho, P. M. (2006). Secondary osteoporosis: a review of the recent evidence. *Endocrine Practice*, *12*(4), 436-445.
15. Tannenbaum, C., Clark, J., Schwartzman, K., Wallenstein, S., Lapinski, R., Meier, D., & Luckey, M. (2002). Yield of laboratory testing to identify secondary contributors to osteoporosis in otherwise healthy women. *The Journal of Clinical Endocrinology & Metabolism*, *87*(10), 4431-4437.

16. Kostenuik, P. J. (2005). Osteoprotegerin and RANKL regulate bone resorption, density, geometry and strength. *Current opinion in pharmacology*, *5*(6), 618-625.
17. Hannan, M. T., Felson, D. T., Dawson-Hughes, B., Tucker, K. L., Cupples, L. A., Wilson, P. W., & Kiel, D. P. (2000). Risk factors for longitudinal bone loss in elderly men and women: the Framingham Osteoporosis Study. *Journal of Bone and Mineral Research*, *15*(4), 710-720.
18. Kanis, J. A., Johansson, H., Johnell, O., Oden, A., De Laet, C., Eisman, J. A., & Tenenhouse, A. (2005). Alcohol intake as a risk factor for fracture. *Osteoporosis international*, *16*(7), 737-742.
19. Hopper, J. L., & Seeman, E. (1994). The bone density of female twins discordant for tobacco use. *New England Journal of Medicine*, *330*(6), 387-392.
20. Kanis, J. A., Johnell, O., Odén, A., Johansson, H., De Laet, C., Eisman, J. A., & Melton, L. J. (2005). Smoking and fracture risk: a meta-analysis. *Osteoporosis International*, *16*(2), 155-162.
21. Schousboe, J. T., Shepherd, J. A., Bilezikian, J. P., & Baim, S. (2013). Executive summary of the 2013 international society for clinical densitometry position development conference on bone densitometry. *Journal of Clinical Densitometry*, *16*(4), 455-466.
22. Gourlay, M. L., Fine, J. P., Preisser, J. S., May, R. C., Li, C., Lui, L. Y., & Ensrud, K. E. (2012). Bone-density testing interval and transition to osteoporosis in older women. *New England Journal of Medicine*, *366*(3), 225-233.
23. Worsfold, M., Powell, D. E., Jones, T. J., & Davie, M. W. (2004). Assessment of urinary bone markers for monitoring treatment of osteoporosis. *Clinical chemistry*, *50*(12), 2263-2270.
24. Delmas, P. D., Eastell, R., Garnero, P., Seibel, M. J., & Stepan, J. (2000). The use of biochemical markers of bone turnover in osteoporosis. *Osteoporosis International*, *11*(18), S2-S17.
25. Eastell, R., Krege, J. H., Chen, P., Glass, E. V., & Reginster, J. Y. (2005). Development of an algorithm for using PINP to monitor treatment of patients with teriparatide*. *Current Medical Research and Opinion®*, *22*(1), 61-66.
26. Bonnick, S. (2004). *Bone Densitometry in Clinical* Practice (2nd ed., pp. 257-63). Totowa, NJ: Humana Press.
27. Kanis, J. A., McCloskey, E. V., Johansson, H., Strom, O., Borgstrom, F., & Odén, A. (2008). Case finding for the management of osteoporosis with FRAX®—assessment and intervention thresholds for the UK. *Osteoporosis international*, *19*(10), 1395-1408.
28. Nguyen, N. D., Frost, S. A., Center, J. R., Eisman, J. A., & Nguyen, T. V. (2008). Development of prognostic nomograms for individualizing 5-year and 10-year fracture risks. *Osteoporosis International*, *19*(10), 1431-1444.
29. Ryan, C. S., Petkov, V. I., & Adler, R. A. (2011). Osteoporosis in men: the value of laboratory testing. *Osteoporosis International*, *22*(6), 1845-1853.
30. Painter, S. E., Kleerekoper, M., & Camacho, P. M. (2006). Secondary osteoporosis: a review of the recent evidence. *Endocrine Practice*, *12*(4), 436-445.
31. Tannenbaum, C., Clark, J., Schwartzman, K., Wallenstein, S., Lapinski, R., Meier, D., & Luckey, M. (2002). Yield of laboratory testing to identify secondary

contributors to osteoporosis in otherwise healthy women. *The Journal of Clinical Endocrinology & Metabolism, 87*(10), 4431-4437.

32. Nicodemus, K. K., & Folsom, A. R. (2001). Type 1 and type 2 diabetes and incident hip fractures in postmenopausal women. *Diabetes care, 24*(7), 1192-1197.

33. Bassett, J. D., O'Shea, P. J., Sriskantharajah, S., Rabier, B., Boyde, A., Howell, P. G., & Samarut, J. (2007). Thyroid hormone excess rather than thyrotropin deficiency induces osteoporosis in hyperthyroidism. *Molecular Endocrinology, 21*(5), 1095-1107.

34. Bauer, D. C., Ettinger, B., Nevitt, M. C., & Stone, K. L. (2001). Risk for fracture in women with low serum levels of thyroid-stimulating hormone. *Annals of Internal Medicine, 134*(7), 561-568.

35. Pasieka, J. L., Parsons, L. L., Demeure, M. J., Wilson, S., Malycha, P., Jones, J., & Krzywda, B. (2002). Patient-based surgical outcome tool demonstrating alleviation of symptoms following parathyroidectomy in patients with primary hyperparathyroidism. *World journal of surgery, 26*(8), 942-949.

36. Boonstra, C. E., & Jackson, C. E. (1971). Serum calcium survey for hyperparathyroidism: results in 50,000 clinic patients. *American journal of clinical pathology, 55*(5), 523-526.

37. Grey, A., Mitnick, M. A., Shapses, S., Ellison, A. Gundberg, C., & Insogna, K. (1996). Circulating levels of interleukin-6 and tumor necrosis factor-alpha are elevated in primary hyperparathyroidism and correlate with markers of bone resorption--a clinical research center study. *The Journal of Clinical Endocrinology & Metabolism, 81*(10), 3450-3454.

38. Bilezikian, J. P., Khan, A. A., & Potts Jr, J. T. (2009). Guidelines for the management of asymptomatic primary hyperparathyroidism: summary statement from the third international workshop. *The Journal of Clinical Endocrinology & Metabolism, 94*(2), 335-339.

39. Giannini, S., Nobile, M., Sella, S., Carbonare, L. D., & Favus, M. J. (2005). Bone disease in primary hypercalciuria. *Critical reviews in clinical laboratory sciences, 42*(3), 229-248.

40. Adams, J. S., Song, C. F., & Kantorovich, V. (1999). Rapid recovery of bone mass in hypercalciuric, osteoporotic men treated with hydrochlorothiazide. *Annals of internal medicine, 130*(8), 658-660.

41. Khosla S, Amin S, Orwoll E (2008) Osteoporosis in men. *Endocr Rev* **29**: 441-64.

42. Stenson, W. F., Newberry, R., Lorenz, R., Baldus, C., & Civitelli, R. (2005). Increased prevalence of celiac disease and need for routine screening among patients with osteoporosis. *Archives of Internal Medicine, 165*(4), 393-399.

43. Sezer, O., Heider, U., Zavrski, I., Kühne, C. A., & Hofbauer, L. C. (2003). RANK ligand and osteoprotegerin in myeloma bone disease. *Blood, 101*(6), 2094-2098.

44. Tian, E., Zhan, F., Walker, R., Rasmussen, E., Ma, Y., Barlogie, B., & Shaughnessy Jr, J. D. (2003). The role of the Wnt-signaling antagonist DKK1 in the development of osteolytic lesions in multiple myeloma. *New England Journal of Medicine, 349*(26), 2483-2494.

45. Barete, S., Assous, N., De Gennes, C., Grandpeix, C., Feger, F., Palmerini, F., & Fraitag, S. (2010). Systemic mastocytosis and bone involvement in a cohort of 75 patients. *Annals of the rheumatic diseases*, *69*(10), 1838-1841.
46. Chiappetta, N., & Gruber, B. (2006, August). The role of mast cells in osteoporosis. In *Seminars in arthritis and rheumatism* (Vol. 36, No. 1, pp. 32-36). WB Saunders.
47. Geisler, J., & Lønning, P. E. (2010). Impact of aromatase inhibitors on bone health in breast cancer patients. *The Journal of steroid biochemistry and molecular biology*, *118*(4), 294-299.
48. Hadji, Peyman. "Aromatase inhibitor-associated bone loss in breast cancer patients is distinct from postmenopausal osteoporosis." *Critical reviews in oncology/hematology* 69.1 (2009): 73-82.
49. Rizzoli, R., Body, J. J., De Censi, A., Reginster, J. Y., Piscitelli, P., & Brandi, M. L. (2012). Guidance for the prevention of bone loss and fractures in postmenopausal women treated with aromatase inhibitors for breast cancer: an ESCEO position paper. *Osteoporosis international*, *23*(11), 2567-2576.
50. Rizzoli, R., Cooper, C., Reginster, J. Y., Abrahamsen, B., Adachi, J. D., Brandi, M. L., & Harvey, N. C. (2012). Antidepressant medications and osteoporosis. *Bone*, *51*(3), 606-613.
51. Rivelli, S. K., & Muzyk, A. J. (2009). Antidepressants and Osteoporosis. *Psychopharm Review*, *44*(8), 57-63.
52. Chung, S., & Ahn, C. (1994). Effects of anti-epileptic drug therapy on bone mineral density in ambulatory epileptic children. *Brain and Development*, *16*(5), 382-385.
53. Sheth, R. D., Wesolowski, C. A., Jacob, J. C., Penney, S., Hobbs, G. R., Riggs, J. E., & Bodensteiner, J. B. (1995). Effect of carbamazepine and valproate on bone mineral density. *The Journal of pediatrics*, *127*(2), 256-262.
54. Nakken, K. O., & Taubøll, E. (2010). Bone loss associated with use of antiepileptic drugs. *Expert opinion on drug safety*, *9*(4), 561-571.
55. Movsowitz, C., Epstein, S., Fallon, M. E. A., Ismail, F., & Thomas, S. (1988). Cyclosporin-A in Vivo Produces Severe Osteopenia in the Rat: Effect of Dose and Duration of Administration*. *Endocrinology*, *123*(5), 2571-2577.
56. Edwards, B. J., Desai, A., Tsai, J., Du, H., Edwards, G. R., Bunta, A. D., & Sprague, S. (2011). Elevated incidence of fractures in solid-organ transplant recipients on glucocorticoid-sparing immunosuppressive regimens. *Journal of osteoporosis*, *2011*.
57. Kaunitz, A. M., Arias, R., & McClung, M. (2008). Bone density recovery after depot medroxyprogesterone acetate injectable contraception use. *Contraception*, *77*(2), 67-76.
58. Meier, C., Brauchli, Y. B., Jick, S. S., Kraenzlin, M. E., & Meier, C. R. (2010). Use of depot medroxyprogesterone acetate and fracture risk. *The Journal of Clinical Endocrinology & Metabolism*, *95*(11), 4909-4916.
59. Thorstenson, A., Bratt, O., Akre, O., Hellborg, H., Holmberg, L., Lambe, M., & Adolfsson, J. (2012). Incidence of fractures causing hospitalisation in prostate cancer patients: results from the population-based PCBaSe Sweden. *European Journal of Cancer*, *48*(11), 1672-1681.

60. Ngamruengphong, S., Leontiadis, G. I., Radhi, S., Dentino, A., & Nugent, K. (2011). Proton pump inhibitors and risk of fracture: a systematic review and meta-analysis of observational studies. *The American journal of gastroenterology, 106*(7), 1209-1218.

61. Lecka-Czernik, B. (2010). Bone loss in diabetes: use of antidiabetic thiazolidinediones and secondary osteoporosis. *Current osteoporosis reports, 8*(4), 178-184.

62. Meier, C., Kraenzlin, M. E., Bodmer, M., Jick, S. S., Jick, H., & Meier, C. R. (2008). Use of thiazolidinediones and fracture risk. *Archives of Internal Medicine, 168*(8), 820-825.

63. Looker, A. C., Orwoll, E. S., Johnston, C. C., Lindsay, R. L., Wahner, H. W., Dunn, W. L., & Heyse, S. P. (1997). Prevalence of low femoral bone density in older US adults from NHANES III. *Journal of Bone and Mineral Research, 12*(11), 1761-1768.

64. Baim, S., & Leslie, W. D. (2012). Assessment of fracture risk. *Current osteoporosis reports, 10*(1), 28-41.

65. Shahinian, V. B., Kuo, Y. F., Freeman, J. L., & Goodwin, J. S. (2005). Risk of fracture after androgen deprivation for prostate cancer. *New England Journal of Medicine, 352*(2), 154-164.

66. Burge, R., Dawson-Hughes, B., Solomon, D. H., Wong, J. B., King, A., & Tosteson, A. (2007). Incidence and economic burden of osteoporosis-related fractures in the United States, 2005–2025. *Journal of bone and mineral research, 22*(3), 465-475.

67. Kiebzak, G. M., Beinart, G. A., Perser, K., Ambrose, C. G., Siff, S. J., & Heggeness, M. H. (2002). Undertreatment of osteoporosis in men with hip fracture. *Archives of Internal Medicine, 162*(19), 2217-2222.

68. Fink, H. A., Ewing, S. K., Ensrud, K. E., Barrett-Connor, E., Taylor, B. C., Cauley, J. A., & Orwoll, E. S. (2006). Association of testosterone and estradiol deficiency with osteoporosis and rapid bone loss in older men. *The Journal of Clinical Endocrinology & Metabolism, 91*(10), 3908-3915.

69. Ryan, C. S., Petkov, V. I., & Adler, R. A. (2011). Osteoporosis in men: the value of laboratory testing. *Osteoporosis International, 22*(6), 1845-1853.

70. US Preventive Services Task Force. (2011). Screening for osteoporosis: US preventive services task force recommendation statement. *Annals of internal medicine, 154*(5), 356.

71. Adler, R. A. (2011). Management of osteoporosis in men on androgen deprivation therapy. *Maturitas, 68*(2), 143-147.

72. Rand, T. H., Seidl, G., Kainberger, F., Resch, A., Hittmair, K., Schneider, B., & Imhof, H. (1997). Impact of spinal degenerative changes on the evaluation of bone mineral density with dual energy X-ray absorptiometry (DXA). *Calcified tissue international, 60*(5), 430-433.

73. Kanis, J. A., Melton, L. 3., Christiansen, C., Johnston, C. C., & Khaltaev, N. (1994). The diagnosis of osteoporosis. *J Bone Miner Res, 9*(8), 1137-1141.

74. Lewiecki, E. M., Watts, N. B., McClung, M. R., Petak, S. M., Bachrach, L. K., Shepherd, J. A., & Downs Jr, R. W. (2004). Official positions of the international

society for clinical densitometry. *The Journal of Clinical Endocrinology & Metabolism, 89*(8), 3651-3655.
75. Watts, N. B., Adler, R. A., Bilezikian, J. P., Drake, M. T., Eastell, R., Orwoll, E. S., & Finkelstein, J. S. (2012). Osteoporosis in men: an Endocrine Society clinical practice guideline. *The Journal of Clinical Endocrinology & Metabolism, 97*(6), 1802-1822.
76. Cummings, S. R., Cosman, F., & Jamal, S. A. (Eds.). (2002). *Osteoporosis: an evidence-based guide to prevention and management*. ACP Press.
77. Frommer, D. J. (1964). Changing age of the menopause. *BMJ, 2*(5405), 349-351.
78. Baxter-Jones, A. D., Faulkner, R. A., Forwood, M. R., Mirwald, R. L., & Bailey, D. A. (2011). Bone mineral accrual from 8 to 30 years of age: an estimation of peak bone mass. *Journal of bone and mineral research, 26*(8), 1729-1739.
79. Cohen, A., Fleischer, J., Freeby, M. J., McMahon, D. J., Irani, D., & Shane, E. (2009). Clinical characteristics and medication use among premenopausal women with osteoporosis and low BMD: the experience of an osteoporosis referral center. *Journal of Women's Health, 18*(1), 79-84.
80. Karlsson, C., Obrant, K. J., & Karlsson, M. (2001). Pregnancy and lactation confer reversible bone loss in humans. *Osteoporosis International, 12*(10), 828-834.
81. Davey, M. R., De Villiers, J. T., Lipschitz, S., & Pettifor, J. M. (2012). Pregnancy-and lactation-associated osteoporosis. *Journal of Endocrinology, Metabolism and Diabetes of South Africa, 17*(3), 149-153.
82. Cheung, A. M., Feig, D. S., Kapral, M., Diaz-Granados, N., & Dodin, S. (2004). Prevention of osteoporosis and osteoporotic fractures in postmenopausal women: recommendation statement from the Canadian Task Force on Preventive Health Care. *Canadian Medical Association Journal, 170*(11), 1665-1667.
83. The Writing Group for the ISCD Position Development Conference. (2004). Diagnosis of osteoporosis in men, premenopausal women, and children. *Journal of Clinical Densitometry, 7*(1), 17-26.
84. Looker, A. C., Orwoll, E. S., Johnston, C. C., Lindsay, R. L., Wahner, H. W., Dunn, W. L., & Heyse, S. P. (1997). Prevalence of low femoral bone density in older US adults from NHANES III. *Journal of Bone and Mineral Research, 12*(11), 1761-1768.
85. Melton, L. J., Chrischilles, E. A., Cooper, C., Lane, A. W., & Riggs, B. L. (2005). How many women have osteoporosis?. *Journal of bone and mineral research, 20*(5), 886-892.
86. Cummings, S. R., & Melton, L. J. (2002). Epidemiology and outcomes of osteoporotic fractures. *The Lancet, 359*(9319), 1761-1767.
87. Schroder, H. M., Petersen, K. K., & Erlandsen, M. (1993). Occurrence and incidence of the second hip fracture. *Clinical orthopaedics and related research, 289*, 166-169.
88. Donaldson, M. G., Cawthon, P. M., Lui, L. Y., Schousboe, J. T., Ensrud, K. E., Taylor, B. C., ... & Black, D. M. (2010). Estimates of the proportion of older white men who would be recommended for pharmacologic treatment by the new US

National Osteoporosis Foundation guidelines. *Journal of Bone and Mineral Research, 25*(7), 1506-1511.
89. Recker, R., Lappe, J., Davies, K., & Heaney, R. (2000). Characterization of perimenopausal bone loss: a prospective study. *Journal of Bone and Mineral Research, 15*(10), 1965-1973.
90. Cosman, F., De Beur, S. J., LeBoff, M. S., Lewiecki, E. M., Tanner, B., Randall, S., & Lindsay, R. (2014). Clinician's guide to prevention and treatment of osteoporosis. *Osteoporosis international, 25*(10), 2359-2381.
91. Díez-Pérez, A., Hooven, F. H., Adachi, J. D., Adami, S., Anderson, F. A., Boonen, S., & Greenspan, S. L. (2011). Regional differences in treatment for osteoporosis. The Global Longitudinal Study of Osteoporosis in Women (GLOW). *Bone, 49*(3), 493-498.
92. Overman, R. A., Yeh, J. Y., & Deal, C. L. (2013). Prevalence of oral glucocorticoid usage in the United States: a general population perspective. *Arthritis care & research, 65*(2), 294-298.
93. Van Staa, T. P., Leufkens, H. G. M., Abenhaim, L., Zhang, B., & Cooper, C. (2000). Use of oral corticosteroids and risk of fractures. *Journal of Bone and Mineral Research, 15*(6), 993-1000.
94. Fraser, L. A., & Adachi, J. D. (2009). Glucocorticoid-induced osteoporosis: treatment update and review. *Therapeutic advances in musculoskeletal disease, 1*(2), 71-85.
95. Canalis, E., Mazziotti, G., Giustina, A., & Bilezikian, J. P. (2007). Glucocorticoid-induced osteoporosis: pathophysiology and therapy. *Osteoporosis International, 18*(10), 1319-1328.
96. Staa, T. V., Staa, T. V., Staa, T. V., Leufkens, H. G. M., & Cooper, C. (2002). The epidemiology of corticosteroid-induced osteoporosis: a meta-analysis. *Osteoporosis International, 13*(10), 777-787.
97. Van Staa, T. P., Leufkens, H. G. M., & Cooper, C. (2001). Use of inhaled corticosteroids and risk of fractures. *Journal of Bone and Mineral Research, 16*(3), 581-588.
98. De Vries, F., Bracke, M., Leufkens, H. G., Lammers, J. W. J., Cooper, C., & Van Staa, T. P. (2007). Fracture risk with intermittent high-dose oral glucocorticoid therapy. *Arthritis & Rheumatism, 56*(1), 208-214.
99. Yao, W., Cheng, Z., Busse, C., Pham, A., Nakamura, M. C., & Lane, N. E. (2008). Glucocorticoid excess in mice results in early activation of osteoclastogenesis and adipogenesis and prolonged suppression of osteogenesis: a longitudinal study of gene expression in bone tissue from glucocorticoid-treated mice. *Arthritis & Rheumatism, 58*(6), 1674-1686.
100. Baron, R., & Rawadi, G. (2007). Targeting the Wnt/β-catenin pathway to regulate bone formation in the adult skeleton. *Endocrinology, 148*(6), 2635-2643.
101. Lane, N. E., Yao, W., Balooch, M., Nalla, R. K., Balooch, G., Habelitz, S., & Bonewald, L. F. (2006). Glucocorticoid-Treated Mice Have Localized Changes in Trabecular Bone Material Properties and Osteocyte Lacunar Size That Are Not Observed in Placebo-Treated or Estrogen-Deficient Mice. *Journal of bone and mineral research, 21*(3), 466-476.

102. Weinstein, R. S., O'Brien, C. A., Almeida, M., Zhao, H., Roberson, P. K., Jilka, R. L., & Manolagas, S. C. (2011). Osteoprotegerin prevents glucocorticoid-induced osteocyte apoptosis in mice. *Endocrinology, 152*(9), 3323-3331.
103. Silverman, S. L., & Lane, N. E. (2009). Glucocorticoid-induced osteoporosis. *Current osteoporosis reports, 7*(1), 23-26.
104. Suzuki, Y., Ichikawa, Y., Saito, E., & Homma, M. (1983). Importance of increased urinary calcium excretion in the development of secondary hyperparathyroidism of patients under glucocorticoid therapy. *Metabolism, 32*(2), 151-156.
105. Baim, S., Binkley, N., Bilezikian, J. P., Kendler, D. L., Hans, D. B., Lewiecki, E. M., & Silverman, S. (2008). Official positions of the International Society for Clinical Densitometry and executive summary of the 2007 ISCD Position Development Conference. *Journal of Clinical Densitometry, 11*(1), 75-91.
106. Homik, J., Cranney, A., Shea, B., Tugwell, P., Wells, G., & Adachi, R. (2000). Bisphosphonates for steroid induced osteoporosis (Cochrane Review) The Cochrane Library, Issue 3, Update Software.
107. Saag, K. G., Shane, E., Boonen, S., Marín, F., Donley, D. W., Taylor, K. A., & Marcus, R. (2007). Teriparatide or Alendronate in glucocorticoid-induced osteoporosis. *New England Journal of Medicine, 357*(20), 2028-2039.
108. MacLean, C., Newberry, S., Maglione, M., McMahon, M., Ranganath, V., Suttorp, M., & Desai, S. B. (2008). Systematic review: comparative effectiveness of treatments to prevent fractures in men and women with low bone density or osteoporosis. *Annals of Internal Medicine, 148*(3), 197-213.
109. (1) Saag KG, Emkey R, Schnitzer TJ, et al. Alendronate for the prevention and treatment of glucocorticoid-induced osteoporosis. Glucocorticoid-Induced Osteoporosis Intervention Study Group. N Engl J Med 1998; 339:292.
110. Reid, D. M., Hughes, R. A., Laan, R. F., Sacco-Gibson, N. A., Wenderoth, D. H., Adami, S.,& Devogelaer, J. P. (2000). Efficacy and Safety of Daily Risedronate in the Treatment of Corticosteroid-Induced Osteoporosis in Men and Women: A Randomized Trial. *Journal of Bone and Mineral Research, 15*(6), 1006-1013.
111. Reid, D. M., Devogelaer, J. P., Saag, K., Roux, C., Lau, C. S., Reginster, J. Y., & Mesenbrink, P. (2009). Zoledronic acid and Risedronate in the prevention and treatment of glucocorticoid-induced osteoporosis (HORIZON): a multicentre, double-blind, double-dummy, randomised controlled trial. *The Lancet, 373*(9671), 1253-1263.
112. Saag, K. G., Shane, E., Boonen, S., Marín, F., Donley, D. W., Taylor, K. A., & Marcus, R. (2007). Teriparatide or Alendronate in glucocorticoid-induced osteoporosis. *New England Journal of Medicine, 357*(20), 2028-2039.
113. Saag, K. G., Zanchetta, J. R., Devogelaer, J. P., Adler, R. A., Eastell, R., See, K., & Warner, M. R. (2009). Effects of teriparatide versus Alendronate for treating glucocorticoid-induced osteoporosis: Thirty-six–month results of a randomized, double-blind, controlled trial. *Arthritis & Rheumatism, 60*(11), 3346-3355.
114. Grossman, J. M., Gordon, R., Ranganath, V. K., Deal, C., Caplan, L., Chen, W., & Volkmann, E. (2010). American College of Rheumatology 2010

recommendations for the prevention and treatment of glucocorticoid-induced osteoporosis. *Arthritis care & research, 62*(11), 1515-1526.

115. Crews, D. C., Plantinga, L. C., Miller 3rd, E. R., Saran, R., Hedgeman, E., Saydah, S. H., & Powe, N. R. (2010). Centers for Disease Control and Prevention Chronic Kidney Disease Surveillance Team. Prevalence of chronic kidney disease in persons with undiagnosed or prehypertension in the United States. *Hypertension, 55*(5), 1102-1109.

116. Moe, S., Drüeke, T., Cunningham, J., Goodman, W., Martin, K., Olgaard, K., & Eknoyan, G. (2006). Definition, evaluation, and classification of renal osteodystrophy: a position statement from Kidney Disease: Improving Global Outcomes (KDIGO). *Kidney international, 69*(11), 1945-1953.

117. Lehmann, G., Ott, U., Kaemmerer, D., Schuetze, J., & Wolf, G. (2008). Bone histomorphometry and biochemical markers of bone turnover in patients with chronic kidney disease Stages 3-5. *Clinical nephrology, 70*(4), 296-305.

118. Coco, M., & Rush, H. (2000). Increased incidence of hip fractures in dialysis patients with low serum parathyroid hormone. *American journal of kidney diseases, 36*(6), 1115-1121.

119. National Osteoporosis Foundation. Osteoporosis and chronic kidney disease updates, 2010
http://www.nof.org/sites/default/files/clinicalupdates/Issue20KidneyDisease/kidney.html

120. Miller, P. D., Jamal, S. A., & West, S. L. (2012). Bone mineral density in chronic kidney disease use and misuse. *Clinical Reviews in Bone and Mineral Metabolism, 10*(3), 163-173.

121. Jamal, S. A., Bauer, D. C., Ensrud, K. E., Cauley, J. A., Hochberg, M., Ishani, A., & Cummings, S. R. (2007). Alendronate treatment in women with normal to severely impaired renal function: an analysis of the fracture intervention trial. *Journal of Bone and Mineral Research, 22*(4), 503-508.

122. Matuszkiewicz-Rowińska, J. (2010). KDIGO clinical practice guidelines for the diagnosis, evaluation, prevention, and treatment of mineral and bone disorders in chronic kidney disease. *Pol Arch Med Wewn, 120*(7-8).

123. Miller, P. D. (2011). The kidney and bisphosphonates. *Bone, 49*(1), 77-81.

124. Hartle JE, Tang X, Kirchner HL, et al. Bisphosphonate therapy, death, and cardiovascular events among female patients with CKD: a retrospective cohort study. Am J Kidney Dis 2012; 59:636.

125. Chlebowski, R. T., Hendrix, S. L., Langer, R. D., Stefanick, M. L., Gass, M., Lane, D., & Khandekar, J. (2003). Influence of estrogen plus progestin on breast cancer and mammography in healthy postmenopausal women: the Women's Health Initiative Randomized Trial. *Jama, 289*(24), 3243-3253.

126. Eastell, R., & Hannon, R. A. (2008). Biomarkers of bone health and osteoporosis risk. *Proceedings of the Nutrition Society, 67*(02), 157-162.

127. Nordin, B. C. (1997). Calcium and osteoporosis. *Nutrition, 13*(7), 664-686.

128. Potts, J. T., & Gardella, T. J. (2011). Parathyroid Hormone and Calcium Homeostasis. *Pediatric Bone: Biology & Diseases*, 109.

129. Straub, D. A. (2007). Calcium supplementation in clinical practice: a review of forms, doses, and indications. *Nutrition in Clinical Practice, 22*(3), 286-296.

130. NIH Concensus Panel (1994). NIH Consensus Conference. Optimal calcium intake. NIH consensus development panel on optimal calcium intake. *JAMA, 272,* 1942-1948.
131. Curhan, G. C., Willett, W. C., Speizer, F. E., Spiegelman, D., & Stampfer, M. J. (1997). Comparison of dietary calcium with supplemental calcium and other nutrients as factors affecting the risk for kidney stones in women. *Annals of Internal Medicine, 126*(7), 497-504.
132. Bolland, M. J., Avenell, A., Baron, J. A., Grey, A., MacLennan, G. S., Gamble, G. D., & Reid, I. R. (2010). Effect of calcium supplements on risk of myocardial infarction and cardiovascular events: meta-analysis. *Bmj, 341,* c3691.
133. Li, K., Kaaks, R., Linseisen, J., & Rohrmann, S. (2012). Associations of dietary calcium intake and calcium supplementation with myocardial infarction and stroke risk and overall cardiovascular mortality in the Heidelberg cohort of the European Prospective Investigation into Cancer and Nutrition study (EPIC-Heidelberg). *Heart, 98*(12), 920-925.
134. Holick, M. F. (2004). Sunlight and vitamin D for bone health and prevention of autoimmune diseases, cancers, and cardiovascular disease. *The American journal of clinical nutrition, 80*(6), 1678S-1688S.
135. Holick, M. F. (2007). Vitamin D deficiency. *New England Journal of Medicine, 357*(3), 266-281.
136. Murad, M. H., Elamin, K. B., Abu Elnour, N. O., Elamin, M. B., Alkatib, A. A., Fatourechi, M. M., & Erwin, P. J. (2011). The effect of vitamin D on falls: a systematic review and meta-analysis. *The Journal of Clinical Endocrinology & Metabolism, 96*(10), 2997-3006.
137. Pfeifer, Michael, et al. "Effects of a short-term vitamin D and calcium supplementation on body sway and secondary hyperparathyroidism in elderly women." *Journal of Bone and Mineral Research* 15.6 (2000): 1113-1118.
138. Gallagher, J. C., Riggs, B. L., Eisman, J., Hamstra, A., Arnaud, S. B., & Deluca, H. F. (1979). Intestinal calcium absorption and serum vitamin D metabolites in normal subjects and osteoporotic patients: effect of age and dietary calcium. *Journal of clinical investigation, 64*(3), 729.
139. Khosla, S. (2001). Minireview: The opg/rankl/rank system. *Endocrinology, 142*(12), 5050-5055.
140. Holick, M. F., Binkley, N. C., Bischoff-Ferrari, H. A., Gordon, C. M., Hanley, D. A., Heaney, R. P., ... & Weaver, C. M. (2011). Evaluation, treatment, and prevention of vitamin D deficiency: an Endocrine Society clinical practice guideline. *The Journal of Clinical Endocrinology & Metabolism, 96*(7), 1911-1930.
141. Forrest, K. Y., & Stuhldreher, W. L. (2011). Prevalence and correlates of vitamin D deficiency in US adults. *Nutrition research, 31*(1), 48-54.
142. Holick, M. F., Biancuzzo, R. M., Chen, T. C., Klein, E. K., Young, A., Bibuld, D., & Tannenbaum, A. D. (2008). Vitamin D2 is as effective as vitamin D3 in maintaining circulating concentrations of 25-hydroxyvitamin D. *The Journal of Clinical Endocrinology & Metabolism, 93*(3), 677-681.

143. Judge, J., Birge, S., & Gloth, F. (2014). Recommendations abstracted from the American Geriatrics Society Consensus Statement on vitamin D for prevention of falls and their consequences. *J Am Geriatr Soc*, *62*(1), 147-52.
144. Tuohimaa, P., Lyakhovich, A., Aksenov, N., Pennanen, P., Syvälä, H., Lou, Y. R., & Manninen, T. (2001). Vitamin D and prostate cancer. *The Journal of steroid biochemistry and molecular biology*, *76*(1), 125-134.
145. Holick, M. F., Binkley, N. C., Bischoff-Ferrari, H. A., Gordon, C. M., Hanley, D. A., Heaney, R. P., ... & Weaver, C. M. (2012). Guidelines for preventing and treating vitamin D deficiency and insufficiency revisited. *The Journal of Clinical Endocrinology & Metabolism*, *97*(4), 1153-1158.
146. Stepan, J. J., Alenfeld, F., Boivin, G., Feyen, J. H., & Lakatos, P. (2003). Mechanisms of action of antiresorptive therapies of postmenopausal osteoporosis. *Endocrine regulations*, *37*(4), 225-238.
147. Jacobsen, D. E., Samson, M. M., Kezic, S., & Verhaar, H. J. J. (2007). Postmenopausal HRT and tibolone in relation to muscle strength and body composition. *Maturitas*, *58*(1), 7-18.
148. Naessen, T., Lindmark, B., & Larsen, H. C. (2007). Hormone therapy and postural balance in elderly women. *Menopause*, *14*(6), 1020-1024.
149. Bea, J. W., Zhao, Q., Cauley, J. A., LaCroix, A. Z., Bassford, T., Lewis, C. E., & Chen, Z. (2011). Effect of hormone therapy on lean body mass, falls, and fractures: Six-year results from the Women's Health Initiative Hormone Trials. *Menopause (New York, NY)*, *18*(1), 44.
150. Bolscher, M., Netelenbos, J. C., Barto, R., & van Buuren, L. M. (1999). Estrogen regulation of intestinal calcium absorption in the intact and ovariectomized adult rat. *Journal of Bone and Mineral Research*, *14*(7), 1197-1202.
151. Lindsay R (2004) Hormones and bone health in postmenopausal women. *Endocrine 24*(3): 223-30.
152. La Vecchia, C., Brinton, L. A., & McTiernan, A. (2002). Cancer risk in menopausal women. *Best Practice & Research Clinical Obstetrics & Gynaecology*, *16*(3), 293-307.
153. Ziel, H. K., & Finkle, W. D. (1975). Increased risk of endometrial carcinoma among users of conjugated estrogens. *New England journal of medicine*, *293*(23), 1167-1170.
154. Grady, D., Gebretsadik, T., Kerlikowske, K., Ernster, V., & Petitti, D. (1995). Hormone replacement therapy and endometrial cancer risk: a meta-analysis. *Obstetrics & Gynecology*, *85*(2), 304-313.
155. Canonico, M., Plu-Bureau, G., Lowe, G. D., & Scarabin, P. Y. (2008). Hormone replacement therapy and risk of venous thromboembolism in postmenopausal women: systematic review and meta-analysis. *Bmj*, *336*(7655), 1227-1231.
156. Cirillo, D. J., Wallace, R. B., Rodabough, R. J., Greenland, P., LaCroix, A. Z., Limacher, M. C., & Larson, J. C. (2005). Effect of estrogen therapy on gallbladder disease. *Jama*, *293*(3), 330-339.

157. Grodstein F, Manson JE, Stampfer MJ, Rexrode K 2008 Postmenopausal hormone therapy and stroke: role of time since menopause and age at initiation of hormone therapy. Arch Intern Med 168:861–866

158. Wells, G., Tugwell, P., Shea, B., Guyatt, G., Peterson, J., Zytaruk, N., & Cranney, A. (2002). V. Meta-analysis of the efficacy of hormone replacement therapy in treating and preventing osteoporosis in postmenopausal women. *Endocrine Reviews*, *23*(4), 529-539.

159. Rossouw JE *et al.* (2004) Risks and benefits of estrogen plus progestin in healthy postmenopausal women: principle results from the Women's Health Initiative randomized controlled trial. *JAMA* **288**(3): 321-33.

160. Torgerson, D. J., & Bell-Syer, S. E. (2001). Hormone replacement therapy and prevention of vertebral fractures: a meta-analysis of randomised trials. *BMC Musculoskeletal Disorders*, *2*(1), 7.

161. Torgerson, D. J., & Bell-Syer, S. E. (2001). Hormone replacement therapy and prevention of nonvertebral fractures: a meta-analysis of randomized trials. *Jama*, *285*(22), 2891-2897.

162. Banks, E., Beral, V., Reeves, G., Balkwill, A., Barnes, I., & Million Women Study Collaborators. (2004). Fracture incidence in relation to the pattern of use of hormone therapy in postmenopausal women. *Jama*, *291*(18), 2212-2220.

163. Nancollas, G. H., Tang, R., Phipps, R. J., Henneman, Z., Gulde, S., Wu, W., & Ebetino, F. H. (2006). Novel insights into actions of bisphosphonates on bone: differences in interactions with hydroxyapatite. *Bone*, *38*(5), 617-627.

164. Miller, P. D. (2011). The kidney and bisphosphonates. *Bone*, *49*(1), 77-81.

165. Rodan, G. A., Seedor, J. G., & Balena, R. (1993). Preclinical pharmacology of Alendronate. *Osteoporosis international*, *3*(3), 7-12.

166. Russell, R. G. G., Watts, N. B., Ebetino, F. H., & Rogers, M. J. (2008). Mechanisms of action of bisphosphonates: similarities and differences and their potential influence on clinical efficacy. *Osteoporosis international*, *19*(6), 733-759.

167. Ebetino, F. H., Hogan, A. M. L., Sun, S., Tsoumpra, M. K., Duan, X., Triffitt, J. T., & Lundy, M. W. (2011). The relationship between the chemistry and biological activity of the bisphosphonates. *Bone*, *49*(1), 20-33.

168. Chavassieux, P. M., Arlot, M. E., Reda, C., Wei, L., Yates, A. J., & Meunier, P. J. (1997). Histomorphometric assessment of the long-term effects of Alendronate on bone quality and remodeling in patients with osteoporosis. *Journal of Clinical Investigation*, *100*(6), 1475.

169. Recker, R. R., Delmas, P. D., Halse, J., Reid, I. R., Boonen, S., García-Hernandez, P. A., & Hu, H. (2008). Effects of intravenous zoledronic acid once yearly on bone remodeling and bone structure. *Journal of Bone and Mineral Research*, *23*(1), 6-16.

170. Roschger, P., Rinnerthaler, S., Yates, J., Rodan, G. A., Fratzl, P., & Klaushofer, K. (2001). Alendronate increases degree and uniformity of mineralization in cancellous bone and decreases the porosity in cortical bone of osteoporotic women. *Bone*, *29*(2), 185-191.

171. Gong, H. S., Song, C. H., Lee, Y. H., Rhee, S. H., Lee, H. J., & Baek, G. H. (2012). Early initiation of bisphosphonate does not affect healing and outcomes of

volar plate fixation of osteoporotic distal radial fractures. *The Journal of Bone & Joint Surgery*, *94*(19), 1729-1736.
172. Li, Y. T., Cai, H. F., & Zhang, Z. L. (2015). Timing of the initiation of bisphosphonates after surgery for fracture healing: a systematic review and meta-analysis of randomized controlled trials. *Osteoporosis International*, *26*(2), 431-441.
173. Hegde, V., Jo, J. E., Andreopoulou, P., & Lane, J. M. (2015). Effect of osteoporosis medications on fracture healing. *Osteoporosis International*, 1-11.
174. Shane, E., Burr, D., Ebeling, P. R., Abrahamsen, B., Adler, R. A., Brown, T. D., & Dempster, D. (2010). Atypical subtrochanteric and diaphyseal femoral fractures: report of a task force of the American Society for Bone and Mineral Research. *Journal of Bone and Mineral Research*, *25*(11), 2267-2294.
175. Schilcher, J., Michaëlsson, K., & Aspenberg, P. (2011). Bisphosphonate use and atypical fractures of the femoral shaft. *New England Journal of Medicine*, *364*(18), 1728-1737.
176. Khosla, S., Burr, D., Cauley, J., Dempster, D. W., Ebeling, P. R., Felsenberg, D., & McCauley, L. K. (2007). Bisphosphonate-associated osteonecrosis of the jaw: report of a task force of the American Society for Bone and Mineral Research. *Journal of Bone and Mineral Research*, *22*(10), 1479-1491.
177. Green, J., Czanner, G., Reeves, G., Watson, J., Wise, L., & Beral, V. (2010). Oral bisphosphonates and risk of cancer of oesophagus, stomach, and colorectum: case-control analysis within a UK primary care cohort. *Bmj*, *341*, c4444.
178. Adami, S., Bhalla, A. K., Dorizzi, R., Montesanti, F., Rosini, S., Salvagno, G., & Cascio, V. L. (1987). The acute-phase response after bisphosphonate administration. *Calcified tissue international*, *41*(6), 326-331.
179. Rosen, C. J., & Brown, S. (2003). Severe hypocalcemia after intravenous bisphosphonate therapy in occult vitamin D deficiency. *New England Journal of Medicine*, *348*(15), 1503-1504.
180. Beitz, J., Ibrahim, A., Scher, N., & Williams, G. (2003). Renal failure with the use of zoledronic acid. *N Engl J Med*, *349*, 1676-1679.
181. Loke, Y. K., Jeevanantham, V., & Singh, S. (2009). Bisphosphonates and atrial fibrillation. *Drug safety*, *32*(3), 219-228.
182. Black, D. M., Cummings, S. R., Karpf, D. B., Cauley, J. A., Thompson, D. E., Nevitt, M. C., & Ott, S. M. (1996). Randomised trial of effect of Alendronate on risk of fracture in women with existing vertebral fractures. *The Lancet*, *348*(9041), 1535-1541.
183. Cummings, S. R., Black, D. M., Thompson, D. E., Applegate, W. B., Barrett-Connor, E., Musliner, T. A., & Vogt, T. (1998). Effect of Alendronate on risk of fracture in women with low bone density but without vertebral fractures: results from the Fracture Intervention Trial. *Jama*, *280*(24), 2077-2082.
184. Black, D. M., Schwartz, A. V., Ensrud, K. E., Cauley, J. A., Levis, S., Quandt, S. A., & Wehren, L. E. (2006). Effects of continuing or stopping Alendronate after 5 years of treatment: the Fracture Intervention Trial Long-term Extension (FLEX): a randomized trial. *Jama*, *296*(24), 2927-2938.
185. Harris, S. T., Watts, N. B., Genant, H. K., McKeever, C. D., Hangartner, T., Keller, M., & Axelrod, D. W. (1999). Effects of Risedronate treatment on

vertebral and nonvertebral fractures in women with postmenopausal osteoporosis: a randomized controlled trial. *Jama*, *282*(14), 1344-1352.

186. Mellström, D. D., Sörensen, O. H., Goemaere, S., Roux, C., Johnson, T. D., & Chines, A. A. (2004). Seven years of treatment with Risedronate in women with postmenopausal osteoporosis. *Calcified tissue international*, *75*(6), 462-468.

187. Watts, N. B., Chines, A., Olszynski, W. P., McKeever, C. D., McClung, M. R., Zhou, X., & Grauer, A. (2008). Fracture risk remains reduced one year after discontinuation of Risedronate. *Osteoporosis international*, *19*(3), 365-372.

188. Grbic, J. T., Landesberg, R., Lin, S. Q., Mesenbrink, P., Reid, I. R., Leung, P. C., & Eriksen, E. F. (2008). Incidence of osteonecrosis of the jaw in women with postmenopausal osteoporosis in the health outcomes and reduced incidence with zoledronic acid once yearly pivotal fracture trial. *The Journal of the American Dental Association*, *139*(1), 32-40.

189. Black, D. M., Reid, I. R., Boonen, S., Bucci-Rechtweg, C., Cauley, J. A., Cosman, F., & Leung, P. C. (2012). The effect of 3 versus 6 years of Zoledronic acid treatment of osteoporosis: A randomized extension to the HORIZON-Pivotal Fracture Trial (PFT). *Journal of bone and mineral research*, *27*(2), 243-254.

190. Kostenuik, P. J. (2005). Osteoprotegerin and RANKL regulate bone resorption, density, geometry and strength. *Current opinion in pharmacology*, *5*(6), 618-625.

191. Prolia (denosumab) (prescribing information). Thousand Oaks, California: Amgen Inc: 2010.

192. Torring, O. (2015). Effects of denosumab on bone density, mass and strength in women with postmenopausal osteoporosis. *Ther Adv Musculoskelet Dis*, 7(3), 88-102.

193. Saith MR, et al. (2009) Denosumab Halt Prostate Cancer Study Group. Denosumab in men receiving androgen-deprivation therapy for prostate cancer. *N Eng J Med* **301**: 745-55.

194. Hu, M. I., Glezerman, I. G., Leboulleux, S., Insogna, K., Gucalp, R., Misiorowski, W., & Jaccard, A. (2014). Denosumab for treatment of hypercalcemia of malignancy. *The Journal of Clinical Endocrinology & Metabolism*, *99*(9), 3144-3152.

195. Kong, Y. Y., Yoshida, H., Sarosi, I., Tan, H. L., Timms, E., Capparelli, C., & Khoo, W. (1999). OPGL is a key regulator of osteoclastogenesis, lymphocyte development and lymph-node organogenesis. *Nature*, *397*(6717), 315-323.

196. Bekker, P. J., Holloway, D. L., Rasmussen, A. S., Murphy, R., Martin, S. W., Leese, P. T., & DePaoli, A. M. (2004). A single-dose placebo-controlled study of AMG 162, a fully human monoclonal antibody to RANKL, in postmenopausal women. *Journal of Bone and Mineral Research*, *19*(7), 1059-1066.

197. Block, G. A., Bone, H. G., Fang, L., Lee, E., & Padhi, D. (2012). A single-dose study of denosumab in patients with various degrees of renal impairment. *Journal of Bone and Mineral Research*, *27*(7), 1471-1479.

198. Bone, H. G., Chapurlat, R., Brandi, M. L., Brown, J. P., Czerwiński, E., Krieg, M. A., & Ivorra, J. A. R. (2013). The effect of three or six years of denosumab exposure in women with postmenopausal osteoporosis: results from the FREEDOM extension.

199. Miller, P. D., Bolognese, M. A., Lewiecki, E. M., McClung, M. R., Ding, B., Austin, M., & San Martin, J. (2008). Effect of denosumab on bone density and turnover in postmenopausal women with low bone mass after long-term continued, discontinued, and restarting of therapy: a randomized blinded phase 2 clinical trial. *Bone*, *43*(2), 222-229.

200. McClung MR, Lewiecki EM, Bolognese MA, *et al.* (2010) 233 8 year abstract. *ISCD* March.

201. Langdahl, B. L., Teglbjærg, C. S., Ho, P. R., Chapurlat, R., Czerwinski, E., Kendler, D. L., & Bolognese, M. A. (2015). A 24-month Study Evaluating the Efficacy and Safety of Denosumab for the Treatment of Men With Low Bone Mineral Density: Results From the ADAMO Trial. *The Journal of Clinical Endocrinology & Metabolism*.

202. Brewer, H. B., Fairwell, T., Ronan, R., Sizemore, G. W., & Arnaud, C. D. (1972). Human parathyroid hormone: amino-acid sequence of the amino-terminal residues 1-34. *Proceedings of the National Academy of Sciences*, *69*(12), 3585-3588.

203. Jilka, R. L., O'Brien, C. A., Ali, A. A., Roberson, P. K., Weinstein, R. S., & Manolagas, S. C. (2009). Intermittent PTH stimulates periosteal bone formation by actions on post-mitotic preosteoblasts. *Bone*, *44*(2), 275-286.

204. Dobnig, H., & Turner, R. T. (1997). The Effects of Programmed Administration of Human Parathyroid Hormone Fragment (1–34) on Bone Histomorphometry and Serum Chemistry in Rats 1. *Endocrinology*, *138*(11), 4607-4612.

205. Lindsay, R., Nieves, J., Formica, C., Henneman, E., Woelfert, L., Shen, V., & Cosman, F. (1997). Randomised controlled study of effect of parathyroid hormone on vertebral-bone mass and fracture incidence among postmenopausal women on oestrogen with osteoporosis. *The Lancet*, *350*(9077), 550-555.

206. Dobnig, H., Sipos, A., Jiang, Y., Fahrleitner-Pammer, A., Ste-Marie, L. G., Gallagher, J. C., Eriksen, E. F. (2005). Early changes in biochemical markers of bone formation correlate with improvements in bone structure during teriparatide therapy. *The Journal of Clinical Endocrinology & Metabolism*, *90*(7), 3970-3977.

207. Borggrefe, J., Graeff, C., Nickelsen, T. N., Marin, F., & Glüer, C. C. (2010). Quantitative computed tomographic assessment of the effects of 24 months of teriparatide treatment on 3D femoral neck bone distribution, geometry, and bone strength: results from the EUROFORS study. *Journal of Bone and Mineral Research*, *25*(3), 472-481.

208. Ma, Y. L., Cain, R. L., Halladay, D. L., Yang, X., Zeng, Q., Miles, R. R., & Onyia, J. E. (2001). Catabolic effects of continuous human PTH (1–38) in vivo is associated with sustained stimulation of RANKL and inhibition of osteoprotegerin and gene-associated bone formation. *Endocrinology*, *142*(9), 4047-4054.

209. Dobnig, H., & Turner, R. T. (1995). Evidence that intermittent treatment with parathyroid hormone increases bone formation in adult rats by activation of bone lining cells. *Endocrinology*, *136*(8), 3632-3638.

210. Keller, H., & Kneissel, M. (2005). SOST is a target gene for PTH in bone. *Bone*, *37*(2), 148-158.

211. Ma, Y. L., Zeng, Q., Donley, D. W., Ste-Marie, L. G., Gallagher, J. C., Dalsky, G. P., & Eriksen, E. F. (2006). Teriparatide increases bone formation in modeling and remodeling osteons and enhances IGF-II immunoreactivity in postmenopausal women with osteoporosis. *Journal of Bone and Mineral Research, 21*(6), 855-864.

212. Cheng, M. L., & Gupta, V. (2012). Teriparatide–indications beyond osteoporosis. *Indian journal of endocrinology and metabolism, 16*(3), 343.

213. Neer, R. M., Arnaud, C. D., Zanchetta, J. R., Prince, R., Gaich, G. A., Reginster, J. Y., & Wang, O. (2001). Effect of parathyroid hormone (1-34) on fractures and bone mineral density in postmenopausal women with osteoporosis. *New England Journal of Medicine, 344*(19), 1434-1441.

214. Thiruchelvam, N., Randhawa, J., Sadiek, H., & Kistangari, G. (2014). Teriparatide Induced Delayed Persistent Hypercalcemia. *Case reports in endocrinology, 2014*.

215. Miller, P. D., Bilezikian, J. P., Diaz-Curiel, M., Chen, P., Marin, F., Krege, J. H., & Marcus, R. (2007). Occurrence of hypercalciuria in patients with osteoporosis treated with teriparatide. *The Journal of Clinical Endocrinology & Metabolism, 92*(9), 3535-3541.

216. Vahle, J. L., Long, G. G., Sandusky, G., Westmore, M., Ma, Y. L., & Sato, M. (2004). Bone neoplasms in F344 rats given teriparatide [rhPTH (1-34)] are dependent on duration of treatment and dose. *Toxicologic pathology, 32*(4), 426-438.

217. Orwoll, E. S., Scheele, W. H., Paul, S., Adami, S., Syversen, U., Diez-Perez, A., & Gaich, G. A. (2003). The effect of teriparatide [human parathyroid hormone (1–34)] therapy on bone density in men with osteoporosis. *Journal of Bone and Mineral Research, 18*(1), 9-17.

218. Saag, K. G., Shane, E., Boonen, S., Marín, F., Donley, D. W., Taylor, K. A., & Marcus, R. (2007). Teriparatide or Alendronate in glucocorticoid-induced osteoporosis. *New England Journal of Medicine, 357*(20), 2028-2039.

219. Krege, J. H., Lane, N. E., Harris, J. M., & Miller, P. D. (2014). PINP as a biological response marker during teriparatide treatment for osteoporosis. *Osteoporosis International, 25*(9), 2159-2171.

220. Black, D. M., Bilezikian, J. P., Ensrud, K. E., Greenspan, S. L., Palermo, L., Hue, T., & Rosen, C. J. (2005). One year of Alendronate after one year of parathyroid hormone (1–84) for osteoporosis. *New England Journal of Medicine, 353*(6), 555-565.

221. Prince, R., Sipos, A., Hossain, A., Syversen, U., Ish-Shalom, S., Marcinowska, E., & Mitlak, B. H. (2005). Sustained nonvertebral fragility fracture risk reduction after discontinuation of teriparatide treatment. *Journal of Bone and Mineral Research, 20*(9), 1507-1513.

222. Bucholz, R. W., MacQueen, M., Rockwood, C. A., & Green, D. P. (2010). *Rockwood and Green's fractures in adults*. Philadelphia, PA: Wolters Kluwer/Lippincott Williams & Wilkins.

223: GEnant H.K., Wu C.Y., van Kuijk C., Nevitt MC (1993). Vertebral fracture assessment using a semiquantitative technique. *Journal of Bone and Mineral Research, 8*(9): 1137-48.

Printed in Great Britain
by Amazon